Horse Owner's Field Guide to Toxic Plants

Horse Owner's Field Guide to Toxic Plants

by Sandra M. Burger and the editors of Breakthrough Publications in consultation with Anthony P. Knight, BVSc, MS, MRCVS

Breakthrough

05 04 03 07 06 05 04

For information address:
Breakthrough Publications Inc.
Ossining, New York 10562

Library of Congress Catalog Card Number: 95-83901
ISBN: 0-914-327-62-3

Book cover and interior designed by Greenboam & Company, Ossining, New York; maps by Robert Greenboam.

Cover photographs, clockwise from left:
Barn—courtesy of Cornell University Plantations;
Squirreltail grass—Glenn Oliver, Visuals Unlimited;
Hydrangeas—David Sieren, Visuals Unlimited;
Common St. Johnswort—John Gerlach, Visuals Unlimited.

The information contained in this book was assembled to assist the horse owner in identifying toxic plants and equine reactions to ingestion of toxins. In all cases of suspected poisoning, it is recommended to consult your veterinarian. This book is not intended to be used as a substitute for veterinary care. The success of the veterinary care administered in cases of suspected poisoning is not under the control of the author or publisher of this guide. Neither Breakthrough Publications nor the authors shall be liable to any person for damage resulting from reliance on information contained in *Horse Owner's Field Guide to Toxic Plants*, including for reasons of any inadvertent error contained herein.

Acknowledgments

My first acknowledgment is to my late grandfather, Edward A. Smith, who always supported and encouraged all of my horse endeavors. I only wish he could have been a part of this one.

Special thanks to my husband Daryl Burger for his patience, encouragement, and understanding. To my mother, Suzanne McQuinn, for her love and support and for editing the draft copy of the manuscript. To my grandmother, Lavera Smith, whose excitement about the project helped me to finish. To the people at St. Peter's Church for their kind words. To my friend Cindy Threadgill and to Pastor Burt Wimberly and his wife Becky. To my mother-in-law, Sue Burger, and to Grandma Saathoff for encouragement. Also thanks to my father, Alan McQuinn, and stepmother, Ramona, who accept me and my horse addiction. I also wish to thank David Burger; Teresa Burger; Ellen, Jim, Ashley, and Curtis Kenalty; Shirley Bell; Kitty Bo Wilson; and all of my family and horse friends. And with great appreciation to Lisa Mooney, my editor at Breakthrough Publications, who took on this project and guided it to publication. For these people who helped with my research: Karen Smith and Sherie Hastings at the Plant Haus in Kerrville, Peggy Crate, Sel Dryden, Corinne Conley, Dr. Quatro Patterson, Dr. Thomerson, Dr. Nightingale, and Dr. Zermuehlen. And lastly, to the inspiration of little Casie Ellebracht and her mare, Blue, who was lost in October 1995, and to Nichole Martin, who lost her show champion, Ellie, in November.

Sandra M. Burger

The publisher wishes to thank Penny O'Prey, project editor; Bob Greenboam, graphic designer; Chris and Dan Diorio, fact checkers and botany researchers; Zoe Moffatt, photo researcher; Chris Heath, proofreader; Judith Hancock, indexer; Peter and Paul Calta, image setters; Rhona Johnson, editor; Michelle Kenneson; Cornell University's Poisonous Plants Garden and Cornell Plantations, Ithaca, New York; the United States Department of Agriculture Photo Archives in Washington, D.C.; Dr. Lynn James, director of the USDA Poisonous Plants Laboratory in Logan, Utah; the Ossining, New York, Public Library and the Westchester County, New York, Library System; the New York Botanical Garden and Library; the Maryknoll Seminary Library, Maryknoll, New York; and Dr. Eleanor Kellon, VMD, for her invaluable advice.

Dedicated
in loving memory
to
Hans

Run free my love,
with the wind
in your beautiful mane.
You were my best friend.
I will miss you always.

Contents

Hans's Story 2

Why Would a Horse Eat a Toxic Plant? 4

Using the *Horse Owner's Field Guide to Toxic Plants* 6

The Plants

- Trees 10
- Bushes, Shrubs, & Vines 34
- Ferns & Palms 74
- Weeds & Wildflowers 78
- Grasses & Horsetails 190

Toxic Suspects 206

Pictorial Glossary 210

Bibliography 212

Photography Credits 213

Index 216

Hans's Story

I am writing this book in the hope that it will prevent others from having to endure the total devastation I have experienced.

I purchased Hans at an auction in Texas when he was just a weanling. His lineage was of working stock horse from South Dakota, and he was a fancy, stout, dun colt. I registered him with the American Quarter Horse Association under the name "I Dun Well" as he was very quick to learn about haltering and standing to be groomed. Due to some medical problems of my own and financial difficulties, I was forced to sell him two months later. It took six years, but I finally purchased him back as an unbroken six-year-old. True to his name, he learned to be ridden very quickly and was much smarter than any other horse I have ever owned.

Hans was my life—my friend and companion. He was no ordinary horse. He was kinder and gentler, more trusting and more beautiful. He liked to play a game with his buckets, making my other horse play tug of war with him as I watched them. He always gave 200 percent when I rode him, and I looked forward to competing on such a fine animal. Hans came to me when I whistled and taught himself to lead without a bridle. He always nickered reassuringly to me when I came home from a long day.

His story is simple. He was fine one day, and off his feed the next. He became colicky that evening, and I took him to our local vet. By morning there was no improvement, so I decided to move him to a hospital that had surgical facilities, 31/2 hours away. There he was diagnosed as having digested poisons, probably the result of eating a toxic plant in his pasture or hay. After blood transfusions, IVs, and numerous drugs to keep him comfortable, we discovered that he was rapidly losing weight and had edema pockets under his belly and a high temperature. The poison was spreading through his entire system.

I drove the long drive almost every day to see him and comfort him. He was all I was able to think about and the stress rendered me unable to work. After two weeks there was some improvement, and he began to nibble at some hay. At last, I was sure he was on the way to recovery! Hans was released to me, and I happily took him home after paying his several-thousand-dollar vet bill.

Two days later, with a temperature of 104 degrees and unable to stand, he lay in his stall quivering and softly nickering to me as I placed cold, wet towels on him to cool him. I rushed him to a well-known racehorse veterinary clinic not far from my home, where he was put on IVs once again and pumped with amino acids and numerous other drugs. He was quarantined, but finally he stabilized to where the IV could be removed. I visited nearly every day and brushed him, taking him from his lonely stall in the quarantine barn to nibble on the grass outside. A week and a half later he was able to eat some grain, and once again I happily picked him up and took him home to see his barn buddy. He ate well the second day after the drive home, although I was cautious not to overfeed him, and once again I was sure we were well on our way to recovery.

On the third day Hans was back off his food, depressed, and not looking well. An apple was all he would accept, so I watched him closely. He went back down and lay on his side, not lifting his head. Back to the clinic we went, where he showed more signs of colic. My veterinarian watched him through the night and called me with bad news in the morning. It was either exploratory surgery, or lose him.

Hans was operated on at 11:30 a.m., May 13, 1994. With two veterinarians working on him, I stood by and watched helplessly. The doctors discovered that he had so many ulcers in his intestine from the poisons and all the medications that he was unable to digest his food properly; he would always be in severe pain. I had no other choice but to agree to have the lethal injection given. It was too late even to say goodbye to my old friend, and now I will never hear his sweet nicker again.

I never found out what plant poisoned Hans, but everyone involved with his case believed he had eaten something that had caused his symptoms. This made me want to learn more about the plants that are poisonous to horses and to thoroughly examine my own pasture as well as the hay I was feeding to my horses. My search led me to my public library and to my local agriculture extension service. I was surprised to learn that there weren't any books specifically for preventing horse poisonings that included both pictures and text written for the lay person. This was the genesis of this book project—I began to see that proper plant identification was the critical first step toward preventing a poisoning.

It is my sincere hope that this book will help save lives.

Sandra M. Burger
Kerrville, Texas
June 30, 1995

Why Would a Horse Eat a Toxic Plant?

Extremely minute amounts of some plants could mean sudden death for a horse. Other plants must be consumed in large quantities over time in order to build up to a toxic level in an animal's system; meanwhile, the horse may be getting poisoned without your knowing it until the levels of toxicity in his or her system are so high that it may be too late for treatment. How can this be avoided? What draws horses to toxic plants in the first place?

• Most toxic plants are unpalatable to horses. Some, however, are quite palatable and appealing to horses, especially in new stages of growth when tender young shoots are sprouting. If you know when a plant is most dangerous and cannot eliminate the plant, you may be able to limit your horse's time on that pasture.

• There may be little else available in the pasture for the horse to eat. Horses are grazing animals, and their digestive systems are built for all-day nibbling. If they are unable to eat healthy foods, they may eat or chew on unusual things to curb their appetites, even poor-tasting toxic plants, if nothing better is available. The best way to avoid this is to prevent overgrazing of a pasture and to supplement the horse's feed with other forage when the grass gets low. Most horses do very well on "free choice" roughage. By setting out a round bale of good quality hay for free choice nibbling, a person can actually prevent his or her horses from being poisoned. Don't forget to put the hay where the weather will not damage it, and put a safety fence around it to prevent animals from lying in it.

• Some poisonous plants may accidentally be mixed with hay or other feed. Horses are frequently fed weedy hay, which may contain low levels of poisonous substances. Animals may not show any reaction to the feed for weeks, months, or even longer. Meanwhile, trouble may be brewing. . . . Suppose, for example, that you are responsible for a horse's maintenance and are feeding cheaper quality hay that, unknown to you, contains toxic weeds. The hay may have aged and developed some dry, yellow spots during storage. Assume that this horse is being worked harder, so you increase the amount of hay you are feeding. The increase in the poor quality hay means an increase in the poisons you are feeding, and weeks later the animal becomes ill from the accumulation of all of these poisons.

For another example, let's say you purchase a new batch of hay that appears to be of much better quality and you feed it to your horse—after all, you buy only the best for this mount! Unknown to you, however, the hay contains a scattering of toxic weeds that are not as well dried as the poorer quality, over-dried hay mentioned above. The "better" hay may have greater potential for poisoning.

• Crops that are used in grain mixtures or hay may have been damaged by frost. This will alter their chemical composition and can actually turn a safe plant into a very harmful one. And what if the feed or hay has become damp from a leaky roof? Do you continue to feed it, even though you can see some mold appearing? As a rule of thumb, if the food doesn't look good, do not feed it to your horse. Hay and feed that have been spoiled by molds can cause colic, founder, or even death. Purchase the best quality feed and hay available, and store it in a dry, well-ventilated area.

• Although processing of forages for hay may remove most of the toxicity in a plant, residues are often left, and while small quantities of such a plant may be safe, eating large amounts can be dangerous. Some plant toxins are cumulative in effect and may be eaten over several months before poisoning becomes evident.

Alkaloids, glycosides, resins, acids, and amines are just a few of the toxic substances that can be found in toxic plants, each requiring specific treatments should your horse ever ingest any of them. Toxicity levels can vary, depending upon the type of soil and climate in which the plants grow, how much water and sunlight they receive, and what stage of growth they have attained when harvested or consumed. Despite a wealth of information available about these factors, scientists still lack knowledge about many plants and what causes their toxicity. Added to the complexity of issues, the age and weight of the horse influence both the effects of the toxins and the treatment for them. In most cases there are no specific treatments. You can see why it is very hard to treat animals that have been poisoned when a person may know neither the quantity of toxin contained in the plant when it was consumed nor even *which* plant was consumed.

Plants do not come with directions and warning labels, yet it is up to us to be sure that our steeds do not come into contact with the wrong ones. If this book should save one horse by making its owner more aware, then the loss of my beloved Hans will not have been for nothing.

Sandra M. Burger

Using the "Horse Owner's Field Guide to Toxic Plants"

This book is intended as a guide for horse owners who want to learn to recognize dangerous plants in the field, hay, or feedbag and thereby prevent a tragedy. Plants that can cause poisoning or mechanical injury to horses are included.

Numbers shown below coincide with the callout numbers on pages 8 and 9.

The book is organized according to types of plants, indicated by color-coded tabs on the left-hand pages ❶:

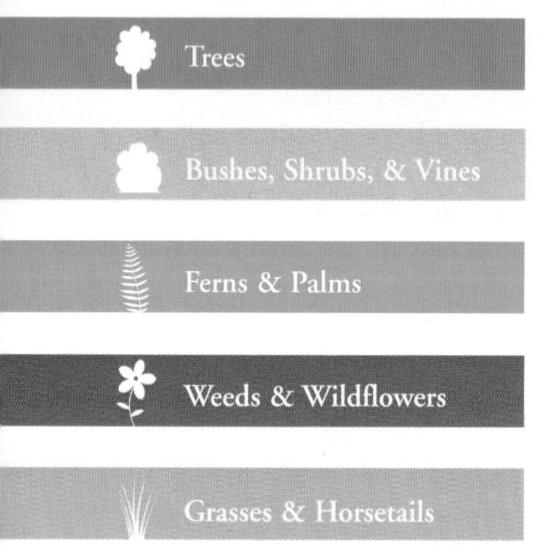

Within these types, the plants are grouped by families ❷, indicated by tabs on the right-hand pages. The families are arranged alphabetically by their scientific names, and within families individual plants are arranged alphabetically by common names. You will notice similar appearances within families.

If you are looking up a plant by name, you can refer to the index at the back of the book for its page number. Another way to look up a plant is to find its category—for instance Trees—and page through that section until you find the plant's name at the top of the left-hand page or a picture that matches it. We believe that visuals are very important in correctly identifying any plant, so we have gathered the best pictures available for each plant in the book. Where possible, close-up photos ❸ and line drawings ❹ show details that can help you identify your plant. As you page through, you may find the visual method the quickest for getting into the correct section of the book.

When you believe you can match the picture in the book to the plant on your property, you will find the following information given for that plant.

Common and scientific names

Some plants are known by different common names ❺, and these may vary in different regions of the country. But each plant has only one scientific name ❻, and the same is true for each plant family. Other names for the plant—and the names of other plants that cause similar reactions—are listed under "Also Known As" and "Similar Species."

Concise description ❼

Size, shape, and color of the various plant parts; the presence of flowers, fruit, seeds, or nuts, and when these appear. (Refer to the Pictorial Glossary at the back of the book for the meanings of terms used in plant descriptions.) A discussion of the poisonous parts of the plant is also included.

Geographic distribution ❽

A general range map accompanies each plant entry. Specific soil and habitat characteristics are also given if appropriate.

Signs of poisoning ❾

The symptoms noted in this book have been documented in horses. Plants that are not toxic to other livestock may actually be quite toxic to horses, and vice versa.

What to do ❿

Practical treatments, if known, and possible prognosis are given. *In all cases of suspected poisoning, it is recommended that you call your veterinarian immediately.*

Keep in mind that *prevention is the best cure.* Whenever you suspect that a plant is dangerous, it is wise to "remove the source from the horse if not the horse from the source." Unfortunately, for many plant poisonings there is no treatment. And because all horses are individuals, there is no way to predict the course or outcome of a poisoning in your particular animal. Therefore, *information in this book must not take the place of immediate veterinary attention* should you suspect your horse has been poisoned. Rather, this book is meant as a reference to help you identify toxic plants—so as to prevent plant poisonings—and to recognize symptoms of poisoning in your horses.

The reason to *call your veterinarian immediately* is that by the time a horse shows signs of poisoning, the condition may be life-threatening. No medications can be given safely until a diagnosis is made. Tranquilizers, pain medication, and other drugs could have harmful effects or even be toxic themselves, depending on the organs already damaged. Intensive care, including IV fluids and cardiac monitoring, may be needed.

While you wait for the vet: Put the horse in a quiet, deeply bedded stall with all water, hay, and feed removed; blanket the horse if it is cold; and keep noise and traffic to a minimum.

❺
❻
❷
❺
Buttercup
❻
Ranunculus species—*Ranunculaceae,* Buttercup Family
❶
Weeds & Wildflowers
Buttercups,
Ranunculus californicus
❸
Buttercup,
Ranunculus bulbosus
Buttercup
Up to 2 feet tall
1
❼
❹

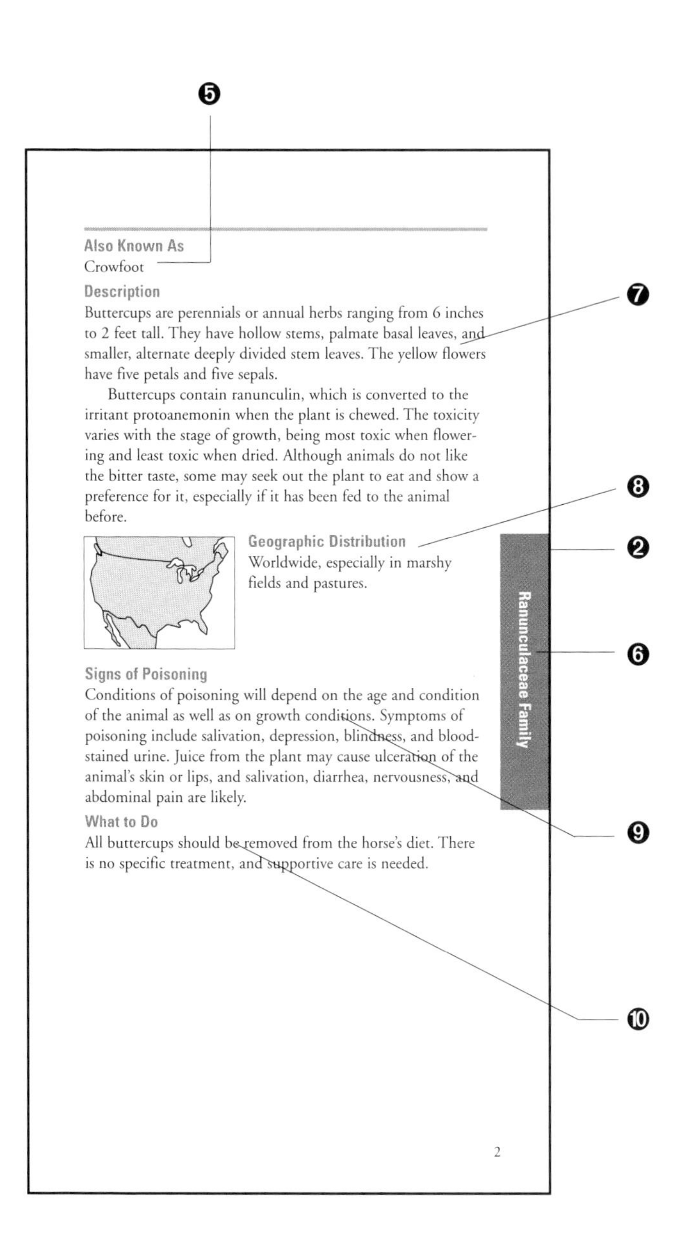

Also Known As
Crowfoot

Description
Buttercups are perennials or annual herbs ranging from 6 inches to 2 feet tall. They have hollow stems, palmate basal leaves, and smaller, alternate deeply divided stem leaves. The yellow flowers have five petals and five sepals.

Buttercups contain ranunculin, which is converted to the irritant protoanemonin when the plant is chewed. The toxicity varies with the stage of growth, being most toxic when flowering and least toxic when dried. Although animals do not like the bitter taste, some may seek out the plant to eat and show a preference for it, especially if it has been fed to the animal before.

Geographic Distribution
Worldwide, especially in marshy fields and pastures.

Signs of Poisoning
Conditions of poisoning will depend on the age and condition of the animal as well as on growth conditions. Symptoms of poisoning include salivation, depression, blindness, and blood-stained urine. Juice from the plant may cause ulceration of the animal's skin or lips, and salivation, diarrhea, nervousness, and abdominal pain are likely.

What to Do
All buttercups should be removed from the horse's diet. There is no specific treatment, and supportive care is needed.

2

Maple, Red

Acer rubrum—*Aceraceae*, Maple Family

Red Maple,
Acer rubrum
Up to 100 feet tall

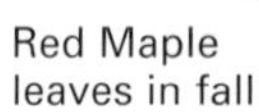

Red Maple
leaves in fall

Description

Red maple grows up to 100 feet tall and has an umbrella shaped top. Its bark is light gray and the wood reddish brown. The palmately shaped leaves have 3 to 5 pointed oval lobes and grow up to 6 inches long in opposite pairs from the dense branches. Leaves are green on top and whitish on the undersides; in fall, they change to shades of red, orange, and yellow. In springtime the tree puts out tiny yellowish red flowers with 5 petals. The tree's winged fruit ripens in spring as well and may fall where horses have access to it.

All red maples are believed toxic to horses, but the toxic principle is yet unknown. Wilted or dried leaves and bark are the most toxic parts, and though horses may not normally eat tree bark, they may nibble on it if tied to a tree and better grazing material is unavailable.

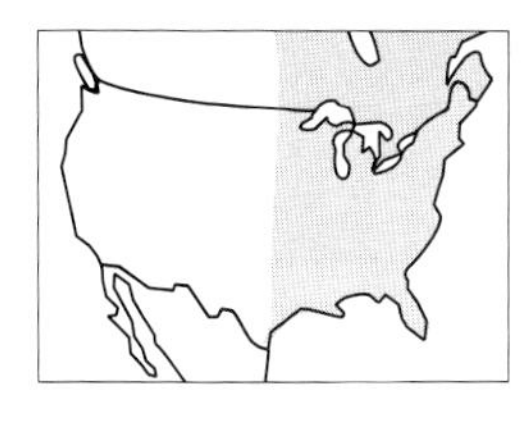

Geographic Distribution

Damp soils in various locations throughout eastern United States and Canada, west to Minnesota and Texas.

Signs of Poisoning

After eating toxic red maple leaves, horses develop severe anemia that results in weakness, depression, pale mucous membranes, and dark brown urine. Death results because red blood cells are unable to transport oxygen to the tissues. Pregnant mares may abort.

What to Do

Call your veterinarian. Fluids, oxygen, and blood transfusions may be helpful if administered early.

Do not plant red maple trees in horse pastures. Always provide a nutritious diet and remove fallen maple branches and leaves from horse pens.

Eve's Necklace

Sophora affinis—*Fabaceae*, Pea Family

Necklace Pod,
Sophora tomentosa
Up to 6–8 feet tall

Necklace Pod,
Sophora tomentosa

Trees

Also Known As

Texas Sophora

Similar Species

Necklace Pod, *Sophora tomentosa*

Description

This tree, often used as an ornamental, grows up to 25 feet tall with an umbrella shaped top. Its branches are slender and brown with green stripes, and the scaly bark on the main stem is brownish red or gray. The pinnately compound, alternate leaves are thin and soft and may be yellow to green on the tops. It flowers in the spring or early summer with pink to white petals. The fruit is a long, curved, black leathery pod containing 1 to 8 seeds (the bean pod of this plant closely resembles that of the poisonous mescal bean). The only parts of the plant to be concerned about are the pods and seeds, as they have been thought to be toxic to livestock.

Geographic Distribution

On hillsides, along waterways, and in limestone beds; in Oklahoma, Louisiana, Texas, and Arkansas.

Signs of Poisoning

Symptoms include loss of coordination, violent trembling, falling, and stiffness, especially when exercising. Respiratory paralysis and death may result in severe cases.

What to Do

No treatment has been noted except to remove the animal from the source, provide supplements and supportive therapy, and keep the animal quiet.

Golden Chain

Laburnum anagyroides—*Fabaceae,* Pea Family

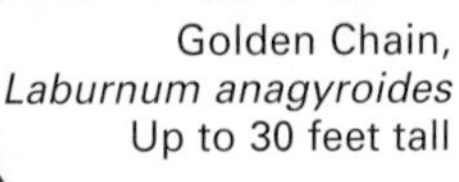

Golden Chain,
Laburnum anagyroides
Up to 30 feet tall

Golden Chain flower cluster

Also Known As

Golden Chain Tree, Bean Tree

Description

This large shrub or small tree grows to 30 feet in height and bears compound leaves of three 3-inch, oblong leaflets. Its 3/4-inch golden-yellow sweetpea-shaped flowers hang in 11/2-foot-long racemes and open in spring. The fruit is a 2-inch-long flattened pod that contains up to eight seeds.

The alkaloid cytisine is the toxic principle present in all parts of the plant, with the seeds being the most poisonous. As little as .05 percent of an animal's weight eaten in seeds will produce toxic symptoms in horses.

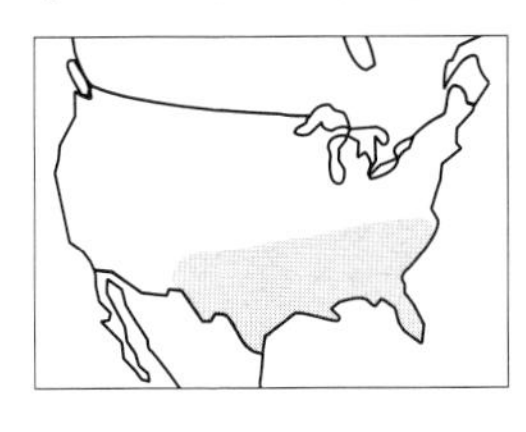

Geographic Distribution

Ornamental, widely cultivated in the southern United States.

Signs of Poisoning

Initial symptoms include excitement, lack of coordination, convulsions, and coma; with eventual death by asphyxiation.

What to Do

No treatment has been noted. Take care not to plant these trees where grazing animals may have access to them.

Locust, Black

Robinia pseudoacacia—*Fabaceae*, Pea Family

Trees

Black Locust,
Robinia pseudoacacia
Up to 75 feet tall

Black Locust,
Robinia pseudoacacia

Black Locust,
Robinia pseudoacacia

Also Known As

False Acacia, Locust Tree

Description

Black locusts are rapidly growing deciduous trees that may grow to 75 feet tall with trunks 30 inches in diameter. Branches are greenish brown, and young ones have sharp spines. The tree has alternate, pinnately compound leaves with more than 10 oval leaflets that grow to 2 inches long and 1 inch wide along an 8- to 14-inch stem. The leaves are green, turning yellow in fall. The trees bloom in May or June with large pealike, 5-petaled flowers that are fragrant and white or cream in color. The fruit of the black locust is a brown, flat pealike pod that ripens in September or October with 4 to 8 seeds.

All parts of the tree are poisonous, containing glycoside robitin and the phytotoxins robin and phasin. Horses have been poisoned by eating the bark while being tied to it, or by stripping the bark from fence posts made from the straight trunks. It takes very little bark to cause poisoning. The phytotoxins are similar in effect to those in Rosary Pea.

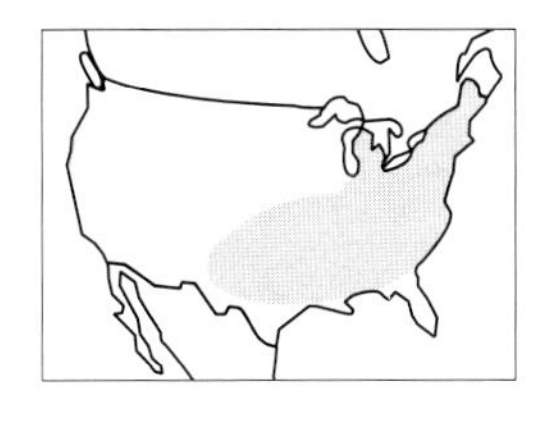

Geographic Distribution

Open woods, roadsides, and pinelands, especially in clay soils; throughout eastern and central United States, especially in the Appalachian Mountains; also at abandoned Texas homesites.

Signs of Poisoning

Poisoned horses will show symptoms of weakness, cold extremities, dilated pupils, weak irregular pulse, diarrhea, stomach pain, and posterior paralysis. Fatal cases are rare.

What to Do

Horses will usually recover if they have not eaten too much of the black locust, but both treatment time and the recovery period are long. See your veterinarian concerning treatment for your particular horse. The administration of activated charcoal, laxatives, and other intensive supportive therapy may aid in recovery.

Oak

Quercus species—*Fagaceae,* Beech Family

Pin Oak, *Quercus* species

White Oak

Oak trees can grow to 150 feet tall

Red Oak

Description

There are more than 60 varieties of oaks. Some species grow to be towering trees—up to 150 feet tall with 3- to 4-foot-thick trunks—while others are shrubs not more than 3 feet tall. Some (live oaks) are evergreen, while others lose their leaves in fall. The leaves of most oaks are elongated and lobed.

All oaks produce acorns containing tannin, and most have gallotannin in their leaves and bark; these parts are the most toxic and can cause severe problems if eaten in great quantities. Young buds also cause problems in springtime. Shin oak (*Q. gambelli*) is the most toxic oak.

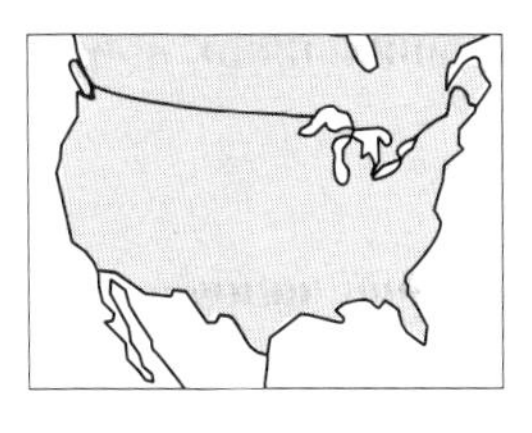

Geographic Distribution

Deciduous woods in diverse conditions throughout the United States and Canada.

Signs of Poisoning

When ingested in great amounts, tannin kills the surface cells of an animal's digestive tract and enters directly into the bloodstream from the gastrointestinal tract, causing anorexia, constipation, rough coat, dry muzzle, abdominal pain, thirst, and frequent urination. Bloody diarrhea develops as a result of intestinal ulceration and necrosis. Kidney and liver damage will be apparent within a week of ingestion and often cause death. Symptoms may last 3 to 10 days, but death occurs about 85 percent of the time unless treatment is administered before the animal consumes much of the plant.

What to Do

Call your veterinarian for treatment information. Calcium hydroxide may be administered orally via nasogastric tube to help prevent further absorption of the poison. Intensive intravenous fluid therapy is necessary to help prevent kidney failure.

Horsechestnut

Aesculus hippocastanum—*Hippocastinaceae,* Buckeye Family

Horsechestnut,
Aesculus hippocastanum
Up to 75 feet tall

Horsechestnut flower

Horsechestnut

Also Known As

Fetid Buckeye, Stinking Buckeye

Similar Species

California, Ohio, Oklahoma, and Texas Buckeyes, *Aesculus* species

Description

Horsechestnut trees can reach 35–75 feet tall, with trunks up to 2 feet in diameter, very thick branches, and brown or gray scaly bark. They have palmately compound, serrated leaves of up to 4–6 inches in length growing opposite each other. Yellowish green or white bell-shaped flowers appear March–May. The leathery, spiny fruit capsules may be seen April–October—depending on the variety—and reach 1- to 11/2-inches long.

The plant may be toxic at any stage of growth, but young shoots and mature fruit are believed to be the most toxic. Young, green sprouts are especially enticing to horses. Toxicity results from a combination of alkaloids, saponins, and especially the glycoside aesculin. As little as 1 percent of an animal's weight in ingested plant material can cause severe poisoning.

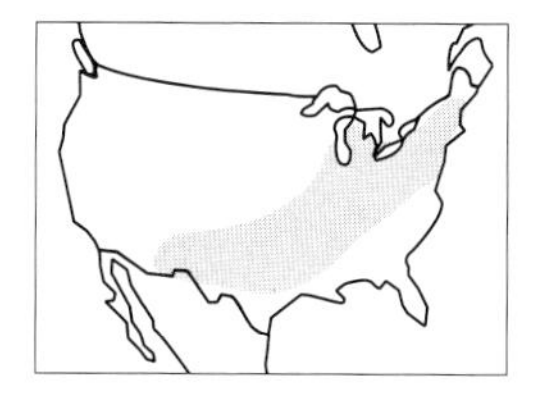

Geographic Distribution

Ornamental and wild; along riverbanks, wet and wooded areas of the Southwest and east to Pennsylvania. Yellow woolly buckeye grows well in the poor soil of Texas Hill Country.

Signs of Poisoning

The toxins affect the nervous system, causing twitching, sluggishness or excitability, and lack of coordination.

What to Do

Stimulants and purgatives may give some relief.

Black Walnut

***Juglans* species**—*Juglandaceae*, Walnut Family

Eastern Black Walnut,
Juglans nigra

Eastern Black Walnut,
Juglans nigra

Black Walnut trees can grow up to 125 feet tall.

Black Walnut flowers

Trees

Similar Species

Eastern Black Walnut, *Juglans nigra;* Texas Black Walnut, *Juglans microcarpa;* Arizona Black Walnut, *Juglans major*

Description

Depending on the variety, the black walnut tree may grow up to 125 feet, with a grayish brown, scaly bark and pinnately compound alternate leaves. It produces spiked flowers in spring or early summer and small oily nuts, which are often harvested, in fall.

The toxic principle juglone, which is a growth inhibitor, is carried in the roots of the tree to other plants surrounding it, inhibiting their growth. Experimentally, juglone has not been shown to cause toxicosis in horses. Furthermore, toxicity is not consistent in all black walnut trees.

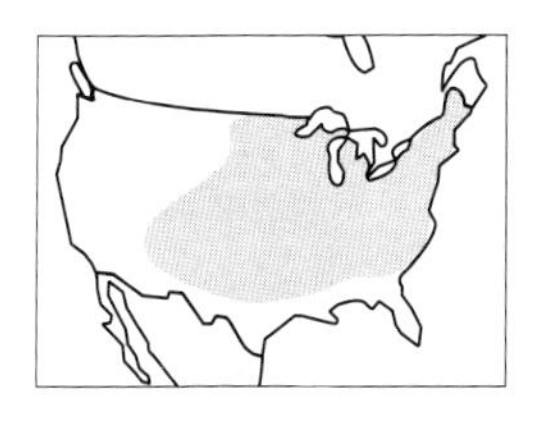

Geographic Distribution

Scattered throughout eastern and midwestern United States, south to Georgia, Texas, and Arizona.

Signs of Poisoning

Horses are affected when the sawdust or shavings from the tree are used for bedding. They show an allergic reaction when standing in the shavings and do not have to ingest them to become sick. As little as 5–20 percent black walnut shavings can cause symptoms of laminitis, swelling in the legs, depression, unwillingness to move, and, in some horses, respiratory difficulties within 12 hours.

What to Do

Remove the horse immediately from the walnut shavings. Wash the horse's legs with mild detergent, and call your veterinarian to treat the laminitis.

Chinaberry

Melia azedarach*—Meliaceae* Family

Trees

Chinaberry
Up to 45 feet tall

Chinaberry branches

Also Known As

Bead Tree, Indian Lilac, Texas Umbrella

Description

The Chinaberry is a fast-growing, umbrella-shaped tree, reaching heights of 45 feet in about 15 years. A hardy species, it has brittle wood and can be recognized by its alternate-growing, pinnately compound leaves, which can reach 2 feet in length at maturity. In spring it has fragrant purple and white flowers of about 1 inch in diameter. Leatherlike, yellowish fruit hangs from the tree in fall and into winter, changing to black berries when they drop; these have been frequently used for rosary beads.

Horses may be attracted to the round cherry-sized berries after they have fallen to the ground. They also may readily eat the berries and bark out of boredom if tied to the tree. The fruit is believed to be the tree's most toxic part, although the entire plant contains toxins and has been used to make insect repellents. Alkaloids and a saponin comprise the toxic principle.

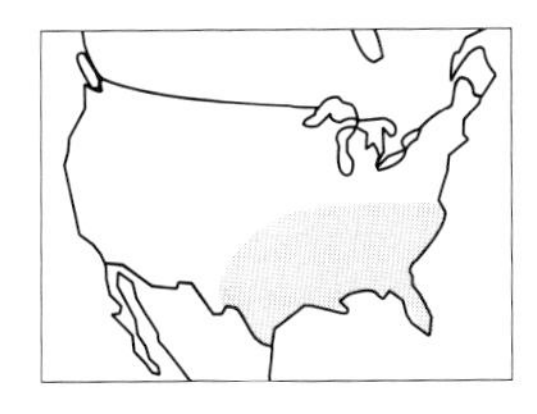

Geographic Distribution

Wild in Texas, Florida, Oklahoma, Arkansas, and North Carolina; cultivated as a bush in fence rows and for decoration.

Signs of Poisoning

Within a few hours of eating the plant, animals become confused and exhibit gastrointestinal colic followed by convulsions.

What to Do

Stimulants, intensive supportive therapy, and cathartics have proven helpful, followed by an easily digestible diet until the animal is fully recovered. Rake up fallen berries if your horse has access to the tree.

Apple

Malus sylvestris—*Rosaceae*, Rose Family

Cultivated apple tree in springtime bloom

Cultivated apple, *Malus domestica*

Apple, *Malus sylvestris*
Up to 50 feet tall

Description

Apples can be poisonous when eaten in large quantities. Although wildlife eat apples with little known effect, horses have been known to develop colic from gorging on fallen apples in their pastures. Cyanide contained in the seeds (at least a cup of seeds) can poison a human or animal.

Apple trees grow to 15–50 feet, depending on the variety. They have wide-spreading branches and a round top. The alternate leaves are usually round and wide and cover the twigs, which are red to brown in color. White or pink 5-petaled flowers may be seen in April or May. The round fruits of 11/3–4 inches in diameter may be green or red, sweet or sour, depending on variety.

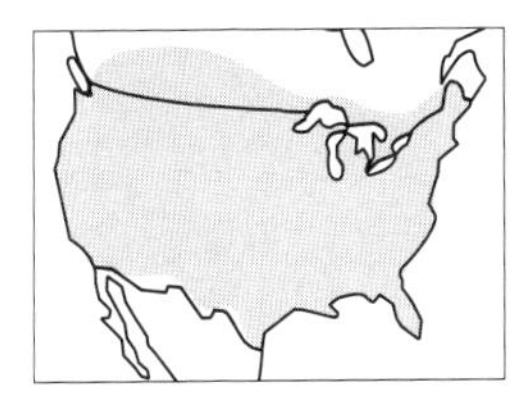

Geographic Distribution

All over the United States and Canada, especially in old fields. Apple trees prefer moderate temperatures and do not grow well in the hot southern climate.

Signs of Poisoning

Symptoms of cyanide poisoning include bright red mucous membranes, anorexia, depression, difficult breathing, weakness, staggering, and death. Horses are more likely to develop colic from overeating apples.

What to Do

Remove horses from the source of apples, especially if there are large numbers of fallen fruits. Mineral oil and other mild laxatives may help remove excess quantities from the intestinal system.

Cherry, Wild

Prunus species—*Rosaceae,* Rose Family

Choke Cherry,
Prunus virginiana

Choke Cherry flowers

Choke Cherry fruits

Choke Cherry
Up to 30 feet tall

Similar Species

Black Cherry, *P. serotina*; Pin Cherry, *P. pensylvanica*; Western Choke Cherry, *Prunus virginiana*

Description

Cherry trees grow up to 30 feet tall or more. They have simple, oblong leaves that grow alternately on reddish brown or orange-brown branches sprouting from a dark, rough trunk. Leaves are darker green on top than on the bottom. Depending on variety, the trees bloom from April through July with 1/4- to 1/3-inch white or pink flowers. The cherries, of similar size, ripen between July and September and are usually black or deep red.

As with apricot and peach trees, the leaves and seeds of the fruit are believed to contain cyanogenetic glycosides. Wild cherry is believed to be the most toxic of these fruiting plants, and its leaves are most toxic when young or wilted.

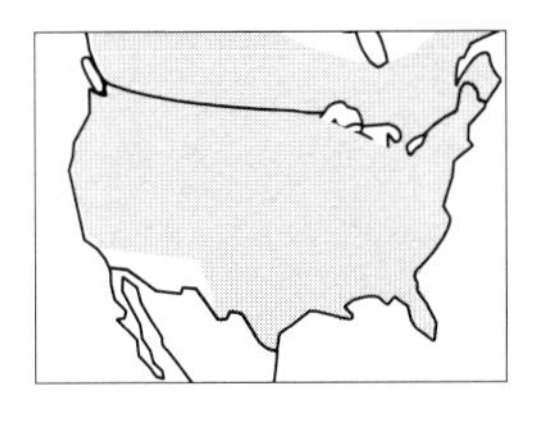

Geographic Distribution

Wild and cultivated throughout Oregon and Washington, southern and eastern United States, and southern Canada; in orchards, prairies, or wooded areas.

Signs of Poisoning

Drinking water shortly after ingestion prompts the quick release of cyanide into the bloodstream, causing slobbering, increased respiration and weak pulse, convulsions, and rapid death. Mucous membranes are bright red.

What to Do

Treatments to counteract the cyanogenic glycosides will need to be administered by your veterinarian immediately. As with other cyanide toxicity, the treatment of choice is a solution of sodium nitrite and sodium thiosulfate administered intravenously. Sedatives and laxatives also may be recommended.

Cherrylaurel

Prunus laurocerasus—*Rosaceae*, Rose Family

Cherrylaurel,
Prunus laurocerasus
Up to 40 feet tall

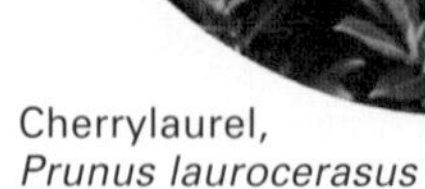

Cherrylaurel,
Prunus laurocerasus

Also Known As
Mock Orange

Similar Species
Carolina Cherry, *Prunus caroliniana*

Description
Cherrylaurel trees grow up to 40 feet tall. Their twigs may be green, red, gray, or brown, and emit a strong cherry odor when crushed, while the main trunk is gray and very rough looking. The leathery evergreen leaves are alternating, simple, and shiny with pointed tips. Small white flowers have 5 petals and are shaped like tiny boats. The unpalatable fruit is 1/2 inch long with thick skin and dry flesh.

Cherrylaurel leaves contain cyanogenic glycosides, which are poisonous to horses and livestock. Seeds can cause cyanide poisoning, similar to wild cherry trees. Birds can apparently feed on the seeds with no ill effects.

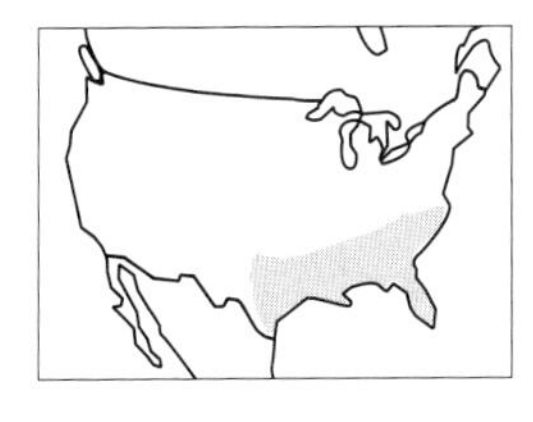

Geographic Distribution
Woods and fence rows in middle and eastern Texas, extending east to Florida, and up into North Carolina. Many people plant them as a bush or hedge in front of their homes.

Signs of Poisoning
Grazing animals have trouble breathing and may suffer bloat, weak heartbeat, staggering, convulsions, coma, and death. Mucous membranes of the mouth and eye will be bright red.

What to Do
Treatments to counteract the cyanogenic glycosides will need to be administered by your veterinarian immediately. As with other cyanide toxicity, the treatment of choice is a solution of sodium nitrite and sodium thiosulfate administered intravenously. Sedatives and laxatives also may be recommended.

Peach

Prunus persica—*Rosaceae*, Rose Family

Trees

Peach,
Prunus persica
Up to 24 feet tall

Peach-tree branch,
Prunus persica

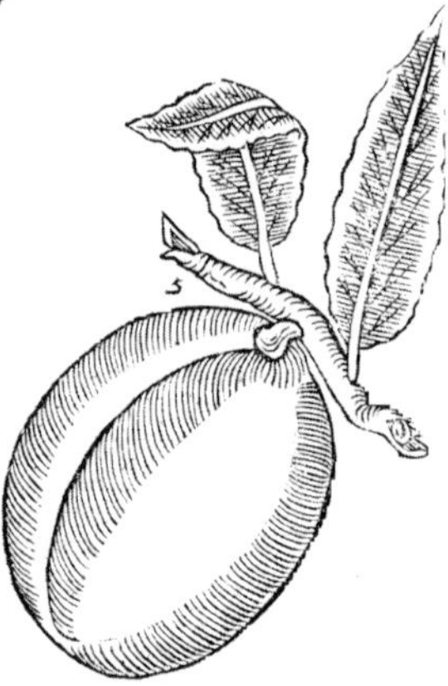

Peach-tree fruit

Description

Peach trees grow to only about 24 feet tall. Numerous 3- to 6-inch, green, simple leaves line the branches. From March to May the trees bloom with 1- to 2-inch-wide flowers having 5 pink petals. Velvet-coated peaches ripen from July to October. All parts of the plant contain cyanide, and the leaves contain prussic acid; the plant is toxic to livestock and poses a potential threat to horses.

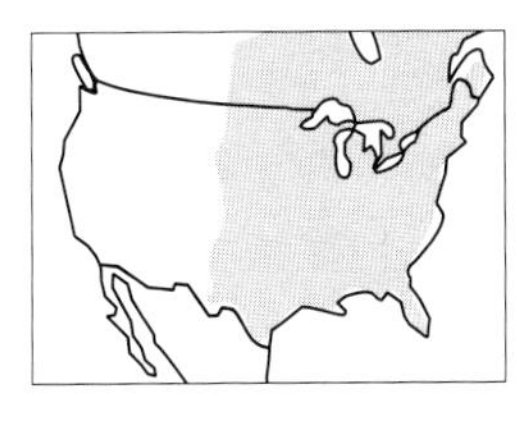

Geographic Distribution

Wild in neglected areas and cultivated from Texas, east and north to Canada.

Signs of Poisoning

Livestock have suffered rapid respiration and gasping; rapid, weak heartbeat; staggering, depression, paralysis, convulsions, coma. Mucous membranes may be bright pink. Death from suffocation may occur quickly.

What to Do

Treatments to counteract the cyanogenic glycosides will need to be administered by your veterinarian immediately. As with other cyanide toxicity, the treatment of choice is a solution of sodium nitrite and sodium thiosulfate administered intravenously. Sedatives and laxatives also may be recommended.

Oleander

Nerium oleander—*Apocynaceae,* Dogbane Family

Oleander,
Nerium oleander
Up to 30 feet tall

Oleander flowers

Bushes, Shrubs, & Vines

Description

Oleander is a woody evergreen bush or small tree growing up to 30 feet tall. It has leathery, pointed leaves that are dark green on the upper sides. These grow opposite each other in groups of three, arranged in whorls, on short stems. Flowers grow in clusters at the ends of the branches and may be white, pink, purplish, or dark red. Fruit grows in a thin, hanging capsule, and seeds are hairy.

Cardiac glycosides, the toxic principle, are found throughout the plant. It takes only 1 ounce of leaves to kill a large horse, or .0005 percent of its body weight. Do not burn this plant, as the smoke is also toxic.

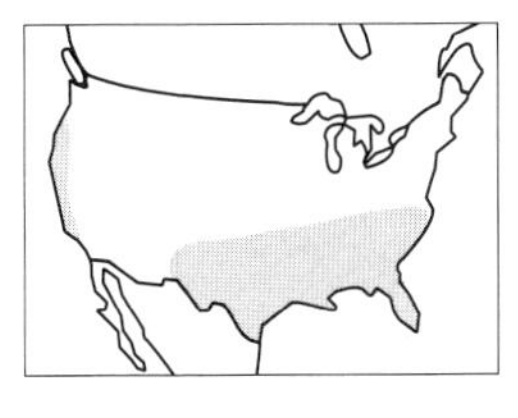

Geographic Distribution

Native to Asia, but cultivated throughout temperate regions of the West Coast and the southern United States.

Signs of Poisoning

In horses, symptoms include diarrhea, trembling, and cold extremities. Paralysis, cardiac arrest, and coma followed by death will occur if a fatal amount is ingested.

What to Do

There is no specific treatment. Activated charcoal via stomach tube, intravenous fluids, and other supportive therapy may be helpful. Specific cardiac drugs may be helpful, but horses will need EKG monitoring in a clinic or hospital setting.

Do not plant oleander in or around areas where horses have access to it. Take caution disposing of oleander prunings.

Boxwood

Buxus sempervirens—*Buxaceae,* Buxus Family

Boxwood,
Buxus sempervirens
Up to 6 feet tall

Boxwood,
Buxus sempervirens

Boxwood

Also Known As

Common Box

Description

The boxwood is a heavily branched bush with leathery leaves that have a dark, glossy upper surface and are generally lighter on the underside. They grow opposite each other and can reach 1/2–11/2 inches long. There are many different varieties, and all are toxic.

Although the toxic alkaloids in this plant are not clearly identified, boxwood has proved highly poisonous to horses and livestock. Only 1 pound of leaves can kill a horse.

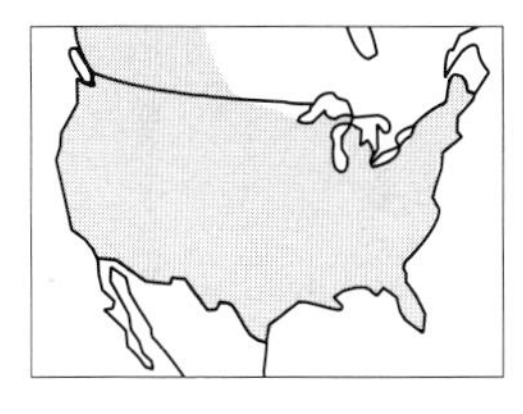

Geographic Distribution

Commonly used for landscaping throughout the United States and western Canada except in the coldest areas.

Signs of Poisoning

Symptoms of poisoning are gastrointestinal distress and blood in the stool. Horses usually die as a result of respiratory failure shortly after exhibiting the initial symptoms.

What to Do

No treatment has been noted. Do not plant boxwood around horse pens or anywhere else horses may have access to it.

Burning Bush

Euonymus atropurpureus—*Celastraceae,* Staff Tree Family

Burning Bush,
Euonymus atropurpureus
Up to 14 feet tall

Burning Bush,
Euonymus atropurpureus

Similar Species

Spindle Tree, *Euonymus europaeus*

Description

Burning bush is a tall shrub or small tree 6–14 feet in height, with greenish gray bark and simple, opposite, small-toothed, oblong leaves. Its twigs and branches are four-angled and its flowers have 4–5 purple petals. Its fruits are smooth capsules, purple-pink in color when ripe in September–October.

The burning bush and spindle tree have been known to cause poisonous reactions after ingestion. Their leaves, bark, and fruit are dangerous to livestock and humans, but the poisonous principle is, as of yet, unknown. Horses are rarely affected.

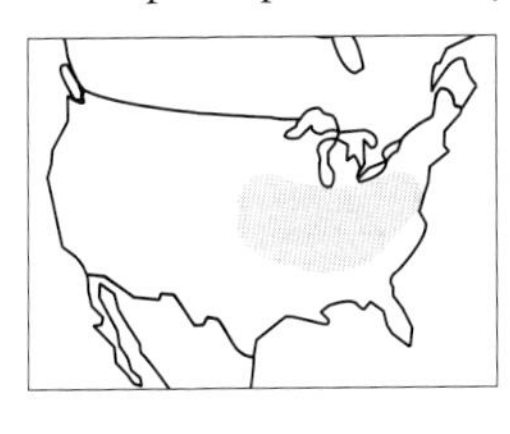

Geographic Distribution

Cultivated, and wild on moist, wooded slopes and stream banks in east and central North America.

Signs of Poisoning

The main symptoms are diarrhea and colic; horses have been noted to experience violent purgation.

What to Do

No treatment has been noted, but symptomatic care should include pain control and replacement of lost fluids and electrolytes.

Do not plant this as an ornamental shrub around horse pens or anywhere horses may have access to it.

Kochia

Kochia scoparia—*Chenopodiaceae,* Goosefoot Family

Kochia,
Kochia scoparia
Up to 5 feet tall

Kochia stems

Also Known As

Mexican Fireweed, Summer Cypress, Burning Brush, Fireball

Description

This is an annual plant that can grow to a 4- or 5-foot height and is frequently used as a landscaping bush around homes. The many-branched stems turn red when the plant matures. Leaves are alternate, thin, and flat and may grow to 2 inches long. Flowers and tiny winged fruit may also be seen.

Kochia is thought to contain alkaloids, soluble oxalates, nitrates, and sulfates. It has also been associated with a substance that destroys thiamine (vitamin B1). The plant is nutritious, however, and can be eaten by cattle and horses without problems under most conditions. Toxicity is probably related to growing conditions; most poisoning appears to occur in late summer, when the plant may be under drought stress.

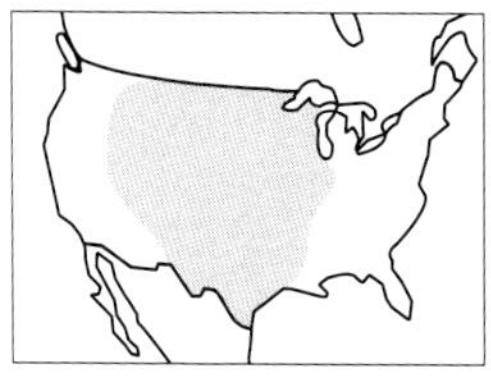

Geographic Distribution

Cultivated; wild in dry soils and neglected areas from Texas, northward.

Signs of Poisoning

Symptoms may include ataxia and muscle spasms, depression, blindness, and photosensitization.

What to Do

Keep photosensitized animals out of the sun, and administer thiamine for blindness and nervous depression.

Saltbush

Atriplex patula—*Chenopodiaceae,* Goosefoot Family

Saltbush,
Atriplex species
Up to 3 feet tall

Saltbush,
Atriplex species

Also Known As

Orach

Similar Species

Four-winged Saltbush, *Atriplex confortifolia*

Description

Saltbush can be recognized by its arrowhead-shaped leaves, 1–3 inches long, growing opposite each other at the bottom of the plant. At the top, the leaves become alternating and silver in color (they are often covered by a whitish powdery substance). Tiny flowers bloom on green spikes, and the seeds are characterized by 4 papery wings (bracts).

Saltbush concentrates selenium and can induce chronic selenium poisoning in horses.

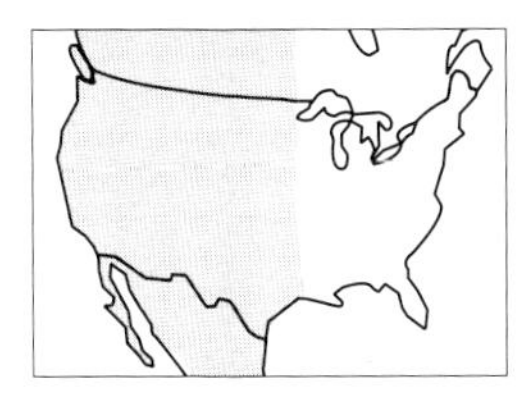

Geographic Distribution

Throughout the western United States, Canada, and Mexico in salt marshes, alkaline soils, and areas that have a high selenium content. In desert areas, the plant is an important forage shrub for livestock.

Signs of Poisoning

Chronic selenium poisoning causes loss of long mane and tail hair, dullness, emaciation, and lameness. Circular ridges with horizontal cracks in all four hooves is also characteristic, and in severe cases, the hoof may separate completely, causing severe lameness.

What to Do

Selenium poisoning is not recognizable until the hair loss and hoof cracking become apparent, and by that time there is no safe means of removing the poison from the horse's system.

Horses raised in high-selenium areas should always be provided with a high-protein diet to ensure sulfur levels adequate to counteract the effects of the selenium. Copper-deficient diets will predispose animals to selenium poisoning, and since most diets are borderline at best for copper, supplementation is advisable. Molasses and bran are good sources of copper.

Azalea, Mock

Menziesia ferruginea—*Ericaceae*, Heath Family

Mock Azalea,
Menziesia ferruginea
Up to 12 feet tall

Mock Azalea flower buds

Mock Azalea

Also Known As

Fool's Huckleberry, Rustyleaf

Description

The mock azalea is a popular deciduous landscaping bush growing 3–12 feet tall. It has alternate, oblong, reddish leaves from 3/4 to 21/2 inches long. It has bell-shaped yellow and red flower clusters and bears a small, woody fruit capsule. Like other members of the heath family, it contains toxic grayanotoxins (andromedotoxin).

Geographic Distribution

Mountains and woods of the northwestern United States and Canada.

Signs of Poisoning

Symptoms after ingestion include nasal discharge, repeated swallowing, salivation, depression, nausea, bloating, colic, teeth grinding, possible collapse, coma, and death. Some liver damage and gastrointestinal irritation has been noted when small amounts of the plants were consumed.

What to Do

Laxatives or activated charcoal via nasogastric tube, if administered promptly, will give the animal some relief and a better chance of survival. Intravenous fluids with appropriate electrolyte replacement is often necessary when severe cardiac toxicity is present.

Labrador Tea, Pacific

***Ledum columbianum*—*Ericaceae*, Heath Family**

Labrador Tea,
Ledum species

Western Labrador Tea,
Ledum glandulosum
Up to 3 feet tall

Similar Species

Western Labrador Tea, *Ledum glandulosum*

Description

These plants are small shrubs, growing only up to 3 feet tall, with yellowish stems and oblong, dotted, 1-inch leaves. The plant also has small white flowers, about 11/2 inch in size, and small fruit.

The toxic principle is grayanotoxin (andromedotoxin).

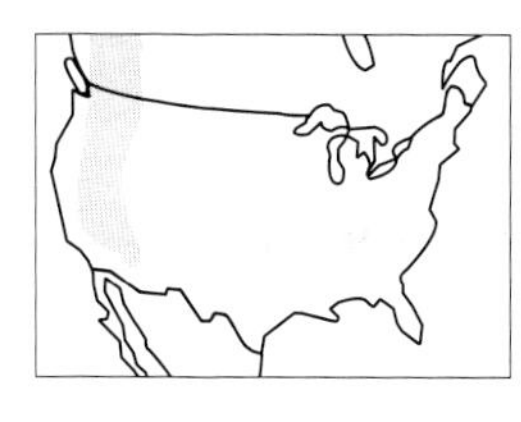

Geographic Distribution

Wet and mountainous areas throughout the western United States and Canada.

Signs of Poisoning

Symptoms after ingestion include nasal discharge, repeated swallowing, salivation, depression, nausea, bloating, colic, teeth grinding, possible collapse, coma, and death. Some liver damage and gastrointestinal irritation has been noted when small amounts of the plants were consumed.

What to Do

Laxatives or activated charcoal via nasogastric tube, if administered promptly, will give the animal some relief and a better chance of survival. Intravenous fluids with appropriate electrolyte replacement is often necessary when severe cardiac toxicity is present.

Laurel, Mountain

Kalmia latifolia—*Ericaceae,* Heath Family

Mountain Laurel, *Kalmia latifolia*

Mountain Laurel, *Kalmia latifolia*

Mountain laurel
Up to 10 feet tall

Bushes, Shrubs, & Vines

Similar Species

Dwarf Laurel, Lambkill, Pale Laurel—*Kalmia* species; Mock Azalea, *Menziesia ferruginea*

Description

The laurels are woody evergreens with 1- to 21/2-inch oblong, leathery leaves that usually grow opposite or whorled in threes. Their showy flowers may be white, pink, or rose. At opposite extremes among the laurels, dwarf laurel grows only to about 10 inches in height, while mountain laurel may reach 10 feet and has alternate leaves. Lambkill grows 1–4 feet tall.

Grayanotoxins (andromedotoxin) are the main toxic components in all laurels and rhododendrons (azaleas) and have similar toxic effects as the cardiac glycosides found in foxglove. Animals may be poisoned within 6 hours of ingestion. These plants cause the most problems in winter and spring when other forage may not be available. Sheep especially have been poisoned by this plant—hence the name lambkill—but horses have rarely been poisoned. Honey made from laurel flowers may be toxic to humans.

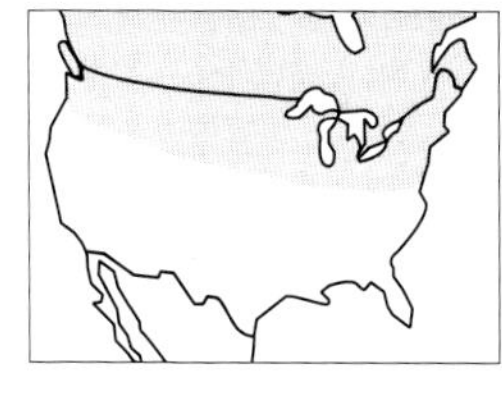

Geographic Distribution

Abandoned pastures, rich wooded areas, rocky soils, and acid bogs in eastern and northern parts of the United States and in Canada.

Signs of Poisoning

Symptoms after ingestion include nasal discharge, repeated swallowing, salivation, depression, nausea, bloating, colic, teeth grinding, possible collapse, coma, and death. Some liver damage and gastrointestinal irritation has been noted when small amounts of the plants were consumed.

What to Do

Laxatives or activated charcoal via nasogastric tube, if administered promptly, will give the animal some relief and a better chance of survival. Intravenous fluids with appropriate electrolyte replacement is often necessary when severe cardiac toxicity is present.

Pieris, Japanese

Pieris japonica—*Ericaceae,* Heath Family

Japanese Pieris,
Pieris japonica
Up to 30 feet tall

Japanese Pieris flower cluster

Similar Species

Pieris floribunda

Description

Japanese pieris is an ornamental, woody, evergreen shrub that grows in a slim formation up to 30 feet tall. It has finely toothed alternating leaves, 11/2–3 inches long, and tiny, 1/4-inch white flowers. It bears a fruit capsule. This plant is closely related to the laurel and rhododendron, and it contains the same toxic principle—grayanotoxin (andromedotoxin).

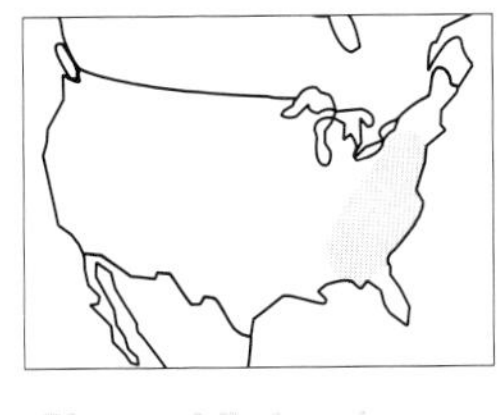

Geographic Distribution

Cultivated widely in landscape gardening; a related species, *Pieris floribunda,* is native to Virginia and Georgia.

Signs of Poisoning

About 6 hours after ingesting just a small amount of the plant, horses may exhibit nasal discharge, repeated swallowing, salivation, depression, nausea, bloating, colic, and teeth grinding. These are often followed by collapse, coma, and death. Animals that survive often have residual liver and kidney damage.

What to Do

Laxatives or activated charcoal via nasogastric tube, if administered promptly, will give the animal some relief and a better chance of survival. Intravenous fluids with appropriate electrolyte replacement is often necessary when severe cardiac toxicity is present.

Rhododendron

Rhododendron maximum—*Ericaceae,* Heath Family

Great Laurel, *Rhododendron maximum*

Rhododendron can grow as a shrub up to 6 feet tall or as a tree up to 35 feet tall.

Rhododendron

Also Known As

Great Laurel, Rosebay

Similar Species

White-Flowered Rhododendron, *Rhododendron albiflorum*

Description

The white-flowered rhododendron is a 3- to 6-foot-tall shrub with thin leaves of only 11/2–3 inches long. Its white 3/4-inch-wide flowers grow in groups from a woody stem, and it has a fruit capsule. The great laurel is an evergreen tree that grows to 35 feet in height. It has alternating, 4- to 10-inch-long, oblong leaves and clusters of mixed rose-pink-white or single-colored flowers, 1–2 inches wide.

Grayanotoxins (andromedotoxin) are the poisonous principle. Animals generally avoid the plant but may eat it if other forage is not available. Poisoning may be fatal as a result of the cardiac effects of the toxin. Horses have rarely been affected.

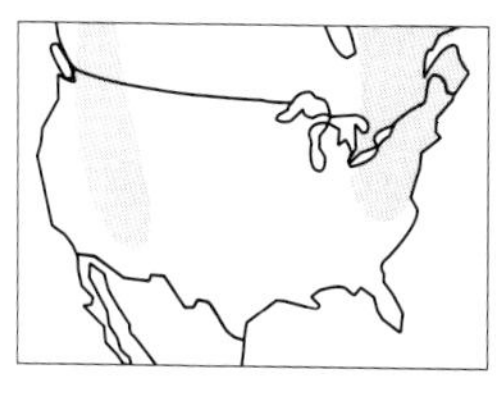

Geographic Distribution

Moist, wooded areas of the eastern United States and Canada. White-flowered rhododendron is seen in wet mountainous areas of the western United States and Canada.

Signs of Poisoning

Horses ingesting the plant may suffer from kidney and liver damage, should they survive. Symptoms include repeated swallowing, salivation, depression, nausea, bloating, abdominal pain (colic), and weakness, followed possibly by coma and then death.

What to Do

Laxatives or activated charcoal via nasogastric tube, if administered promptly, will give the animal some relief and a better chance of survival. Intravenous fluids with appropriate electrolyte replacement is often necessary when severe cardiac toxicity is present.

Castor Bean

Ricinus communis—*Euphorbiaceae*, Spurge Family

Castor Bean, *Ricinus communis* Up to 15–40 feet tall

Castor Bean spiny fruit cluster

Castor Bean

Description

The castor bean grows up to 40 feet tall in tropical climates and to about 15 feet elsewhere. The leaves are shaped like starfish (palmately lobed) and may be 1–2 feet wide, with 6 to 11 rough-edged lobes. From August to September it has maturing fruits that are round, soft, spiny, and 1/2–1 inch in diameter.

The castor bean has been used for its medicinal benefits (castor oil), but the seeds contain the active poison ricin, and merely .01 percent of a horse's weight in seeds can be fatal (ricin is not present in castor oil). Poisonings have generally occurred when castor beans have become mixed with grain at the processing plant or when harvested in the field with forage crops. Horses dislike the taste, but they may take a mouthful, mistaking it for their regular ration, and only one mouthful could contain enough poison to kill them.

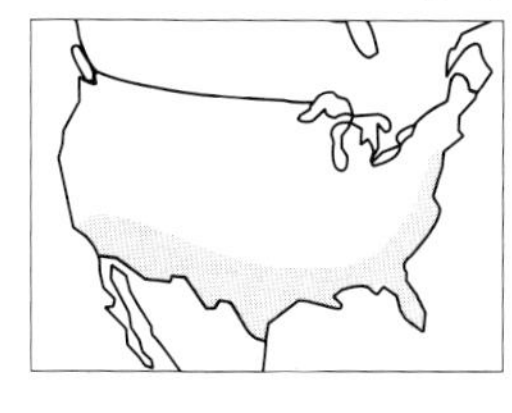

Geographic Distribution

Wild in neglected and disturbed areas of southern California, Texas, east to Florida, and north to New Jersey; cultivated throughout the southern United States as an ornamental.

Signs of Poisoning

Symptoms may appear several hours or even days after ingestion. They include hemorrhaging in the stomach, intestines, and bladder; colic; diarrhea, decreased urination, jaundice; weakness; trembling, sweating, increase of pulse, convulsions; and lack of coordination. Prognosis is very poor.

What to Do

Specific antiserums—determined by the presence of seeds and a specialized blood test and administered by a veterinarian—are the treatment of choice; sedatives and arecoline hydrobromide followed by saline purgatives may also help. Activated charcoal, supportive intravenous fluids, and anticonvulsant therapy administered early in the course of poisoning may increase chances of survival. Ascorbic acid has been shown to increase survival rates.

Do not plant castor beans where horses can gain access to them.

Mescal Bean

Sophora secundiflora—*Fabaceae,* Pea Family

Texas Mountain Laurel, *Sophora secundiflora*

Texas Mountain Laurel flower cluster

Texas Mountain Laurel / Mescal Bean can grow as a shrub to 10 feet tall or as a tree to 35 feet tall.

Also Known As

Texas Mountain Laurel, Frijolito, Carol Bean

Description

The mescal bean is an evergreen shrub, growing to 10 feet, or small tree, growing to 35 feet. It has pinnately compound, alternate growing leaves with shiny leaflets. Its violet to blue flowers hang in showy clusters and are very fragrant. The bright red seeds grow in a dark, woody pod similar to that of Eve's necklace.

Cytisine, a quinolizidine alkaloid, is the toxic principle. All parts of the plant are poisonous, with the seeds being the most toxic, especially when crushed. Horses, however, seem to have a higher tolerance to the plant than livestock and have rarely been affected.

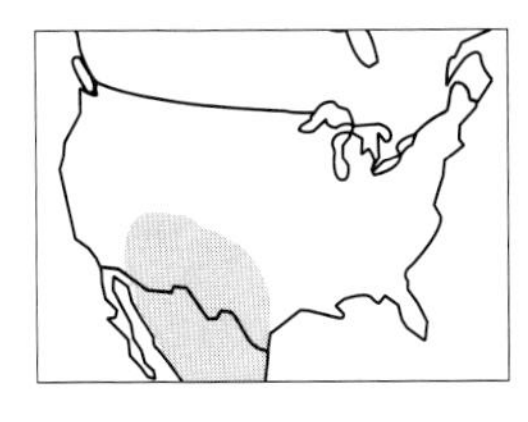

Geographic Distribution

Hills, canyons, limestone soils; in southwestern United States and Mexico.

Signs of Poisoning

Symptoms include loss of coordination, violent trembling, falling, and stiffness, especially when exercised. Respiratory paralysis and death may result in severe cases.

What to Do

There is no treatment except to remove the animal from the source, provide supplements and supportive therapy, and keep the animal quiet.

Mesquite

Prosopis glandulosa*—Fabaceae,* Pea Family

Velvet Mesquite,
Prosopis velutina

Mesquite leaves,
Prosopis juliflora

Mesquite can grow up to 30 feet tall.

Mesquite

Also Known As

Honey Mesquite

Description

Mesquite is a thorny shrub or tree growing to a maximum of 30 feet, with alternate leaves and yellowish green spiked flowers. Its seed pods are brown, oblong, and flattened between the seeds; its rough bark is reddish brown or gray depending on the variety.

The toxic principle in mesquite is thought to be arabinose, and the plant presents the greatest dangers in summer and fall.

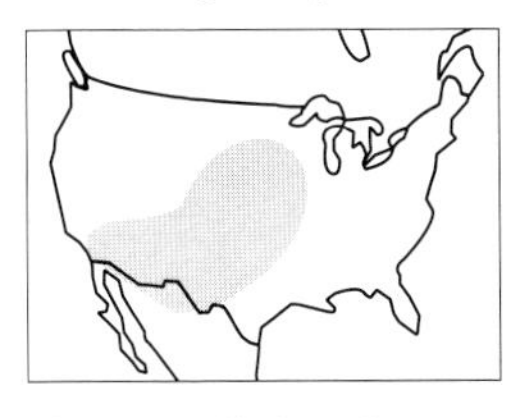

Geographic Distribution

Midwestern and southwestern United States; on dry ranges and in draws.

Signs of Poisoning

Horses that consume the plant may show signs of colic resulting from the seeds/pods, causing impactions. Other symptoms include salivation, continuous chewing, nervousness, tremors, and anemia. A horse's intestines may also suffer severe damage from the thorns of the mesquite.

What to Do

There is no known antidote to the toxin. Colic symptoms must be treated, and colic surgery may be necessary if impaction of the intestinal tract has occurred.

Rosary Pea

Abrus precatorius—*Fabaceae,* Pea Family

Rosary Pea,
Abrus precatorius
Climbs up to
20 feet high

Rosary Pea
seed pod and seeds

Bushes, Shrubs, & Vines

Also Known As

Crabs-Eye, Precatory Bean, Jequirity Bean

Description

This hairy climbing vine attaches to whatever may be in its way, to a height of 10–20 feet. The rosary pea has green or gray stems; pinnately compound, alternating leaves; and many small red or purple flowers. The seed pods are often collected, and the little red seeds, each with a black spot, are extracted and used as rosaries and necklaces.

The seeds contain the highly poisonous phytotoxin, abrin. Just one, if chewed thoroughly, can kill a human or animal. Only 1/2 ounce of powdered seeds can produce symptoms in an adult horse and 2 ounces are lethal. Animals can build resistance to abrin if fed tiny amounts of the substance daily for a number of months.

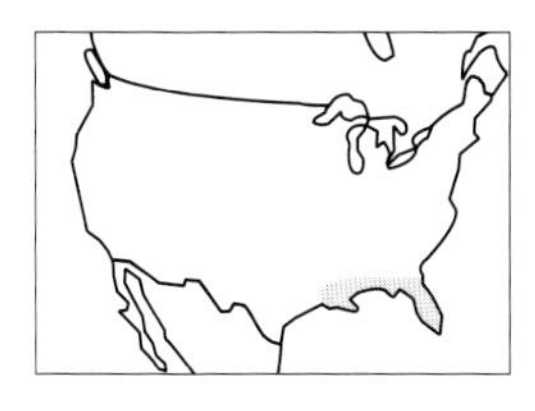

Geographic Distribution

Central and southern Florida; roadsides, thickets, along fencerows, and often in citrus groves.

Signs of Poisoning

The symptoms include gastrointestinal irritation, loss of appetite, high and then low temperatures, and lack of coordination, followed by lung congestion, ulceration of the digestive tract, and kidney, liver, and bladder disorder.

What to Do

It is inadvisable to allow animals any access to the plants or seeds. Supportive therapy will be necessary to counteract the toxic effects of the phytotoxin.

Singletary Pea

***Lathyrus* species**—*Fabaceae,* Pea Family

Sweetpea,
Lathyrus odoratus
Up to 6 feet long

Sweetpea,
Lathyrus odoratus

Sweet Pea

Similar Species

Sweetpea, *Lathyrus odoratus*

Description

Both the singletary pea and the sweetpea are similar vining herbs with pinnately compound leaves accompanied by tendrils. They have white, pink, red, or blue flowers and oblong pods.

Both plants are poisonous to horses and livestock when large amounts are consumed, the seeds being particularly toxic. The toxic principle is beta aminopropionitrile.

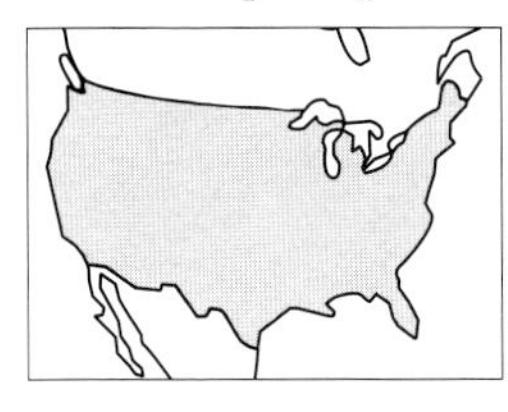

Geographic Distribution

Wild and cultivated throughout the United States, often as a winter forage (and sweetpea as an ornamental).

Signs of Poisoning

After the plant is consumed in quantity over a period of days or weeks, horses exhibit painful attempts at walking and have a hopping gait suggestive of laminitis. Muscle stiffness resembles that seen in horses that are "tied-up."

What to Do

There is no specific treatment, but intensive supportive care may help less severely affected horses recover.

Jessamine, Yellow

Gelsemium sempervirens—*Loganiaceae,* Logania Family

Carolina Jasmine,
Gelsemium sempervirens
Climbs up to 20 feet high

Carolina Jasmine,
Gelsemium sempervirens

Bushes, Shrubs, & Vines

Also Known As

Carolina Jasmine, Carolina Wild Woodbine

Description

Yellow jessamine is a perennial evergreen vine that produces fragrant yellow flowers from February to April. The stem is reddish or green and grows sparsely up to 20 feet, often climbing other trees, fences, or whatever may be in its way. The 2-inch leaves grow opposite one another on the branches, and the flowers, 1–1 1/2 inches long, grow in clusters. The pointed fruit capsule is somewhat flat and oblong.

Winter and spring are the dangerous times for poisoning. Yellow jessamine contains numerous strychnine-related alkaloids such as gelsemine and gelseminine, toxic to horses as well as livestock. The roots and flower nectar are most toxic.

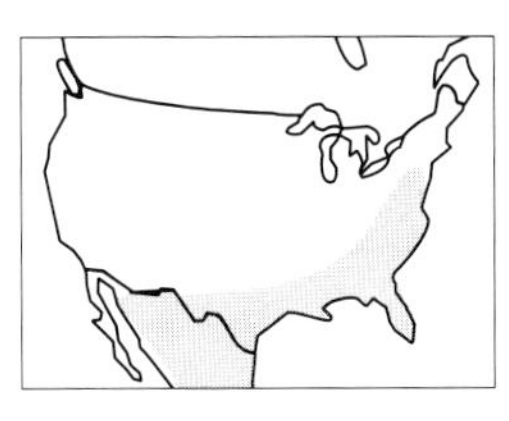

Geographic Distribution

Throughout the eastern and southern United States and Mexico, in dry to wet wooded areas and thickets.

Signs of Poisoning

Poisoning causes weakness, convulsions, rigid extremities, coma, respiratory failure, and death.

What to Do

There is no specific treatment, but animals can be successfully treated with activated charcoal, saline cathartics, intravenous fluids, and supportive care.

Do not plant this species where animals can reach it.

Privet

Ligustrum vulgare—*Oleaceae* Family

Vicary Golden Privet,
Ligustrum vicaryi
Up to 12 feet tall

Privet flowers
Ligustrum vulgare

Privet leaf
arrangement

Also Known As

Ligustrum

Description

The common privet, seen most often in hedges, has dark green, hairless oval leaves that grow opposite one another. Paler on the underside, the leaves are 1–21/2 inches long with a waxy coating. The shrub produces clusters of small fragrant white flowers in spring. Its blue-black oval berries grow in clusters to 1/4 inch in size.

Both berries and leaves are toxic. Their poisonous principles are the glycosides ligustrin and ligustron. Horses and livestock rarely eat the plant.

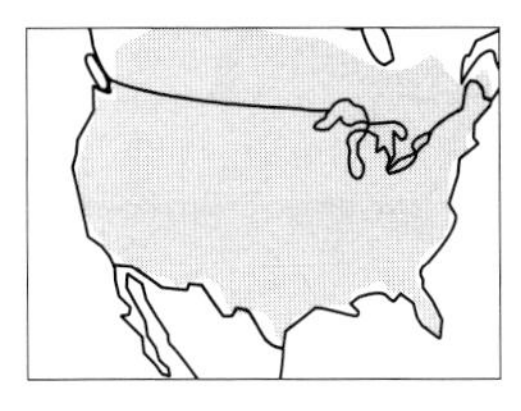

Geographic Distribution

A landscaping favorite throughout the United States and Canada.

Signs of Poisoning

Horses show symptoms of diarrhea, severe colic, lack of coordination, weak pulse, and convulsions shortly after ingestion of a small quantity of the fruit. Death may occur.

What to Do

Give supportive and symptomatic treatment and correct dehydration. Do not plant privets where animals can have easy access to them.

Hydrangea

Hydrangea species—*Saxifragaceae,* Saxifrage Family

Big-leafed Hydrangea, *Hydrangea macrophylla*

Hydrangea, *Hydrangea* species

White Hydrangea, *Hydrangea* species

Hydrangea
Up to 5 feet tall

Description

Hydrangea is a woody shrub 3–5 feet in height, with 3- to 6-inch-long, broad, oval, opposite-branched, toothed leaves. It has masses of flowers in dense heads, varying from white to pink and blue, depending on the acidity of the soil.

The toxic principle is a cyanogenetic glycoside, possibly along with other toxins. Cyanide poisoning may occur if animals are allowed free access to hydrangeas grown as ornamentals or to their prunings.

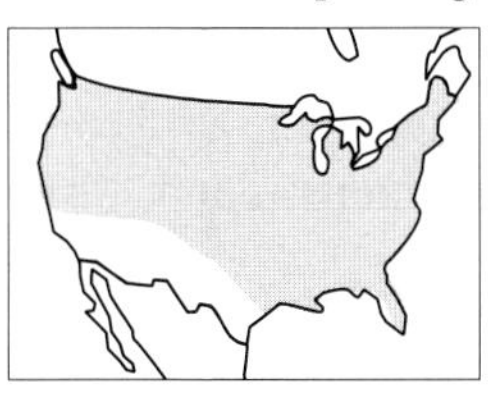

Geographic Distribution

Common garden plant, in moist soils throughout the United States but especially in the South.

Signs of Poisoning

Gastroenteritis and bloody diarrhea will be apparent after consumption. Labored breathing, weakness, coma, and death will occur when toxic levels of cyanide are absorbed from the intestinal tract.

What to Do

Treatment will need to be administered by your veterinarian immediately. As with other cyanide toxicity, the treatment of choice is a solution of sodium nitrite and sodium thiosulfate administered intravenously. Horses may be treated for colic symptoms and diarrhea. Sedatives also may be recommended.

Keep all potted plants and garden plants away from horses and other livestock.

Ground Hemlock

Taxus canadensis—*Taxaceae*, Yew Family

Japanese Yew,
Taxus cuspidata
Up to 5 feet tall

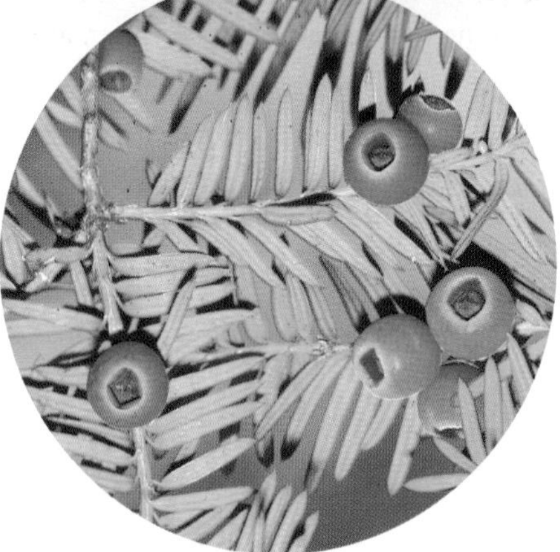

Japanese Yew fruit,
Taxus cuspidata

Similar Species

Japanese Yew, *Taxus cuspidata*; Western or Pacific Yew, *Taxus brevifolia*; English Yew, *Taxus baccata*

Description

Ground hemlock, or American yew, grows only up to 5 feet tall, but western yew can reach 75 feet. All yews have linear, short spiny needles and red-brown scaly bark. Their fruit is bright red at maturity and has a single stone.

The evergreen leaves, fruit, and seeds are toxic whether fresh or dried in hay, and the Japanese yew is reported to cause death in horses. Taxine, an alkaloid, is the poisonous principle at work. Animals will readily eat these plants even if other forage is present. Horses will show signs of poisoning after consuming only about .1 percent of their body weight of green forage.

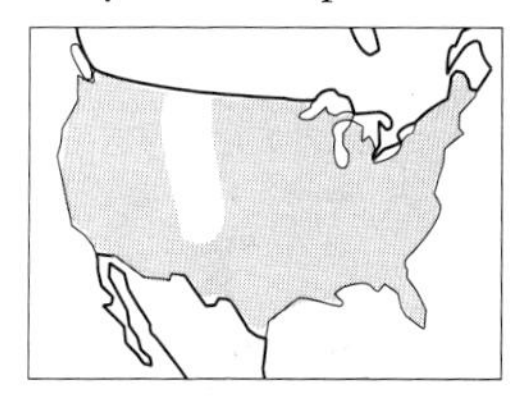

Geographic Distribution

In nature and as ornamentals throughout the United States. Japanese yew is common in the north; western yew grows in the Northwest; English yew grows in mild climates.

Signs of Poisoning

Some horses will react immediately and collapse beside the offending plant, while others may show symptoms later, after full digestion of the plant. The heart will slow and circulation fails; other signs before collapse include nervousness or confusion, diarrhea, and irritation to the digestive tract. Death is the usual outcome.

What to Do

Early use of atropine may help counter the effects of the alkaloid. Oral activated charcoal and saline cathartics along with intravenous fluids as needed are indicated. Artificial respiration may be needed.

s-39-Lantana

Lantana camara—*Verbenaceae*, Verbena Family

Lantana, *Lantana* species

Lantana flowers

Lantana
Up to 6 feet tall

Description

Lantana is a small shrub of irregular shape, with spreading branches that grow up to 6 feet tall. Leaves, which may grow as large as 43/4 inches long and 21/2 inches wide, are firm and oblong. They have a rough surface with prominent veins, grow in whorls of three, and have a strong odor when crushed. In most varieties of lantana, the stems are prickly and, when broken, have a bad odor. The tubular flowers grow in clusters as large as 2 inches across at the top of the plant. These may be different shades of white, yellow, pink, orange, red or purple. The round fleshy fruits, approximately 1/4 inch in size, are dark green, turning black when ripe.

"Lantadene A and B" are the poisonous principles, and ornamental lantana is more toxic than the wild variety. The plant is generally avoided because of its odor, but may be consumed when other forage is scarce.

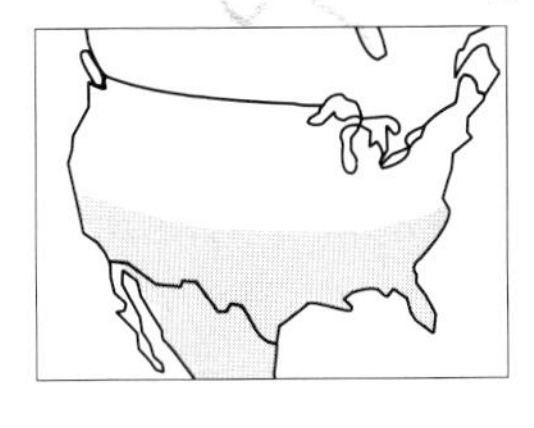

Geographic Distribution

Sandy soils of the southern U.S. coastal plains and Central America.

Verbenaceae Family

Signs of Poisoning

After ingestion of leaves or unripe berries, anorexia, weakness, constipation, and gastroenteritis develop. Gall bladder inflammation and liver disease may occur after larger quantities of lantana have been eaten. Photosensitization may develop secondarily to the liver disease, and death may occur when liver damage is severe.

What to Do

There is no specific treatment. Immediately remove the animal from the source and consult your veterinarian. Photosensitized animals should be kept out of the sun.

Fern Palm

Cycas circinalis—*Cycadaceae*, Cycad Family

Fern Palm,
Cycas revoluta

Some species grow up to 15 feet tall.

Fern Palm,
Cycas revoluta

Similar Species

Florida Arrowroot, *Zamia integrifolia*; Sago Palm, *Cycas* species

Description

These plants resemble ferns, with pinnately compound palm-like leaves growing from thick pithy stems.

Cycasin, a carcinogenic glycoside, can be found throughout the plant. Though the toxin level is highest in the seeds, the foliage often affects animals. Horses have not been reported to be affected, although there is potential for their poisoning.

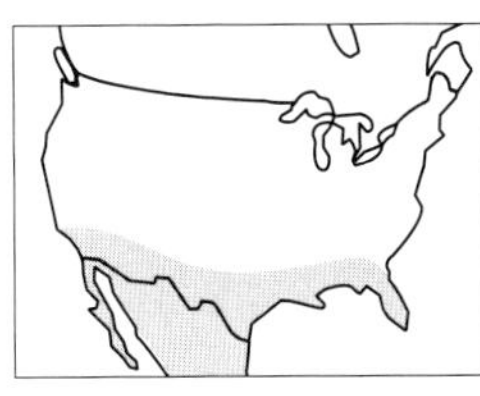

Geographic Distribution

In sand dunes and wooded areas of tropical climates and throughout the southern hemisphere.

Signs of Poisoning

The toxin causes anemia, depression, diarrhea, gastroenteritis, hemorrhaging, nausea, coma, paralysis, and possible death. Liver or kidney damage and tumors may also be apparent.

What to Do

There is no treatment for this poisoning except to control the hemorrhages, treat the anemia, and give other intensive supportive treatment.

Bracken Fern

Pteridium aquilinum—*Dennstaedtiaceae,* Fern Family

Bracken fern,
Pteridium aquilinum

Bracken fern,
Pteridium aquilinum

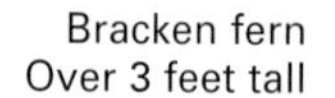

Bracken fern
Over 3 feet tall

Also Known As

Eagle Fern, Hog Brake

Description

This plant has a dark horizontal stalk that grows at least 3 feet tall, with segments of leaves triangular in shape. The leaves are firm, leathery fronds. Brown spore-bearing structures are produced on the undersides and along the edges of the leaves.

Thiaminase is the toxic substance found in this plant that affects horses. Horses rarely eat bracken fern unless other forages are scarce. Cattle and sheep are also affected by bracken fern but as a result of other toxic components.

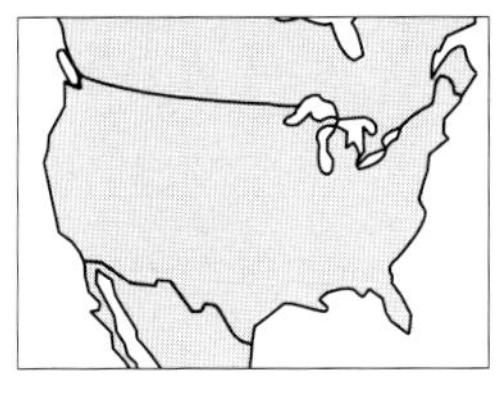

Geographic Distribution

Worldwide; in most forested areas, overgrown fields; poor or sandy soils.

Signs of Poisoning

After eating the plant for a month or more, horses become uncoordinated, depressed, and blind and stand in a braced position with legs apart. At first, appetite may not diminish, even when other symptoms occur. Animals usually die within several days to weeks if they are not removed from the source.

What to Do

Bracken fern poisoning is highly treatable if diagnosed in time. If thiamine deficiency is detected early enough (specialized blood tests can do this), giving megadoses of thiamine and other supportive therapy is effective. A nutritious diet should be given to the horse to prevent resumption of eating the plant.

Atamasco Lily

Zephyranthes atamasco—Amaryllidaceae, Amaryllis Family

Atamasco Lily,
Zephyranthes atamasco
Up to 18 inches tall

Atamasco Lily,
Zephyranthes atamasco

Also Known As

Easter Lily, Rain Lily, Zephyr Lily

Description

The atamasco lily looks and grows much like an onion. It has flat, very narrow, bluish leaves growing up to 14 inches long from its underground bulb. The flower is white with 6 petals and short yellow stamens; it is usually up to 4 inches wide and may have a pale pink stripe on its petals. Single flowers are produced on leafless hairy stems.

The toxic principle is unknown. Animals will consume the lily when other forage is unavailable, usually in late fall and early spring.

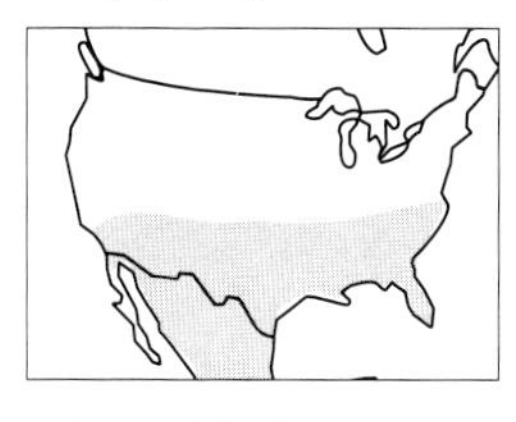

Geographic Distribution

Moist fields or wooded areas; southern United States and Mexico. The plant generally emerges a few days after a heavy rain.

Signs of Poisoning

Symptoms of poisoning include staggering, diarrhea, collapse, and then death. This lily is also known to cause sand burn, a photosensitivity condition.

What to Do

No treatment has been noted except to remove the animal from the infected field and treat symptomatically.

Poison Hemlock

Conium maculatum—*Apiaceae*, Carrot Family

Poison Hemlock,
Conium maculatum
Up to 10 feet tall

Poison Hemlock
in bloom

Poison Hemlock

Weeds & Wildflowers

Also Known As

Spotted Hemlock, European Hemlock, Nebraska or California Fern

Description

Poison hemlock has a fernlike appearance, growing up to 10 feet tall with a purple-spotted stem and a singular white taproot. Its lacy, triangular leaves resemble those of a carrot and have a musky odor, like that of a parsnip, when crushed. It is a biennial with small white flowers clustered in flat-topped umbels and grayish round fruit with ridges, produced in the second year.

A combination of toxic coniine and other alkaloids make this plant nearly as deadly as water hemlock, depending upon its growth stage. The root is least poisonous, with toxicity increasing in leaves and stems in the second year, and the highest concentration in the seeds. Horses and livestock may find the plant's texture palatable and can suffer severe poisoning.

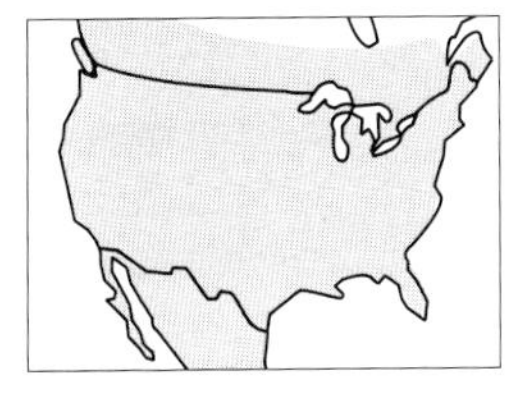

Geographic Distribution

Along roadsides, field edges, and neglected areas; throughout the United States, southern Canada, and into Mexico.

Signs of Poisoning

Bloating, nervousness, trembling, pupil dilation, weakened heart beat, cold extremities, paralysis, coma, respiratory failure, and finally death, can be seen within a few hours or several days of ingestion of the plant.

What to Do

There is no specific antidote for hemlock poisoning. Orally administered mineral oil, activated charcoal, and saline cathartics may help evacuate the gastrointestinal tract. Supportive and intensive care should be given as needed.

Poison hemlock should be pulled up from all areas where horses have access to it.

Water Hemlock

Cicuta maculata—*Apiaceae*, Carrot Family

Water Hemlock,
Cicuta maculata
Up to 8 feet tall

Water Hemlock flower umbel

Weeds & Wildflowers

Also Known As

Musquash Root, Spotted Cowbane

Description

Water hemlock is a perennial herb with a stout, sometimes purple-spotted stalk growing 2–8 feet tall. The stem is swollen at the base where it emerges from a cluster of tubers that are part of the root. Its leaves look like those of dill and may grow to 1 foot or more in length, and its tiny green and white flowers grow in umbels, umbrella-shaped clusters. A yellowish oil (highly toxic) can be seen when the stem is cut.

Members of this genus are considered the most deadly of all plants; their toxic principle is cicutoxin. One mouthful can kill an animal within 15 minutes. The roots, young leaves, and stems are the most toxic parts of the plant, the flowers and seeds being less so. Water hemlock poisoning is most common in the springtime when the enticing green shoots first appear.

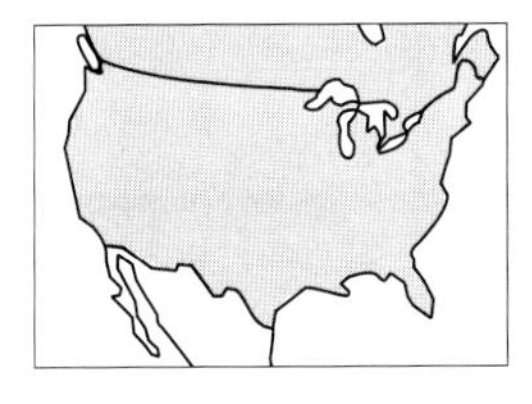

Geographic Distribution

Boggy areas throughout the United States and Canada.

Signs of Poisoning

The cicutoxin causes violent convulsions within half an hour after ingestion. Salivation, tremors, grinding of teeth, dilation of the pupils, elevated temperature, abdominal pain, and bloat are all symptoms of poisoning. Horses may go into convulsions and die within minutes of consumption. Some may last as long as 8 hours and suffer a very painful death.

What to Do

Sedatives are recommended to help with the pain and heart action. Intestinal evacuation may help.

Dogbane

Apocynum cannabinum—*Apocynaceae,* Dogbane Family

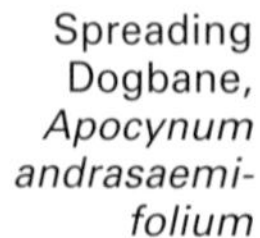

Spreading Dogbane, *Apocynum andrasaemi-folium*

Spreading Dogbane

Spreading Dogbane, *Apocynum andrasaemifolium*

Dogbane
Up to 5 feet tall

Also Known As

Indian Hemp

Description

Dogbane is similar to milkweed with its milky juice, opposite-paired leaves, and long narrow pods hanging in pairs. It can grow to 5 feet tall and has long, narrow seeds with tufts of silky white hairs.

Dogbane contains alkaloids and cardiac glycosides, and horses have died from eating merely 1/2 ounce of this highly toxic plant. But it has a bitter taste so is generally avoided unless other forage is scarce.

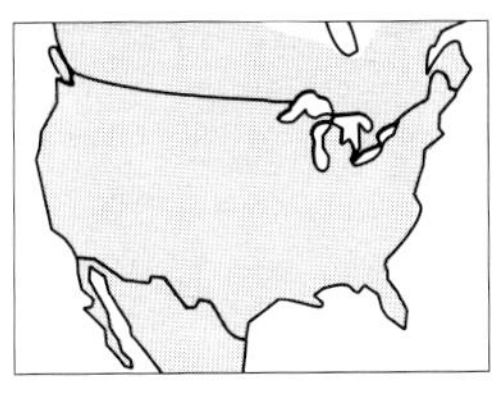

Geographic Distribution

Along streams and in open places; throughout the United States and into Canada and Mexico.

Signs of Poisoning

Symptoms include bloating, weak and rapid pulse, increased temperature, staggering, convulsions, and death.

What to Do

Treatments to counteract the cardiac glycosides will need to be administered by your veterinarian immediately. Horses will need EKG monitoring in a clinic or hospital setting along with specific cardiac drugs. Sedatives, laxatives, and IV fluids also may be recommended.

Milkweed

Asclepias species—*Asclepiadaceae,* Milkweed Family

Milkweed, *Asclepias* species

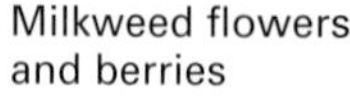

Milkweed flowers and berries

Milkweed
Up to 4 feet tall

Description

These perennials are named for the milklike substance they release when cut. Milkweed leaves are arranged in whorls or in opposite pairs. Small five-petaled flowers grow in clublike formations, forming showy rounded tops on the thick green stems. The flowers may be white, pink, or orange but are usually white, and the leaves will be stiff. The characteristic pods are filled with seeds bearing silky white hairs that aid in wind distribution and propagation.

Of 36 species of milkweed, the narrow-leafed varieties are more toxic than the broad-leafed species. Milkweeds contain galitoxin resinoid, which has irritant properties, but the toxic principle is the cardiac glycosides found in all parts of the plant. The plant is toxic even in dried form.

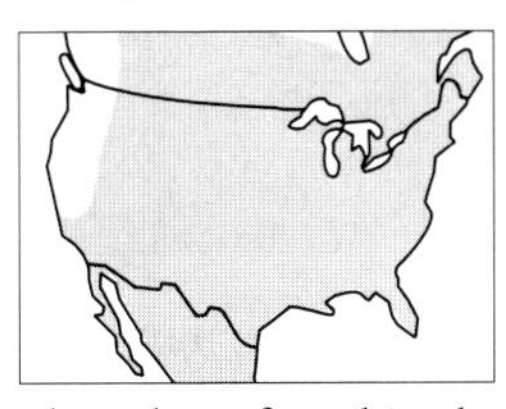

Geographic Distribution

Dry soils, neglected places, overgrazed areas, or fields where it may be cut and bailed into hay. Some plants growing in the eastern United States are approximately 100 times less toxic than those found in the western states.

Signs of Poisoning

Bloating, staggering, rapid pulse, gastroenteritis, depression, weakness, seizures, high temperature, and labored breathing may preceed death. Symptoms appear within a few hours of ingesting the plant.

What to Do

Treat for gastrointestinal distress and cardiac glycoside intoxication: sedatives, laxatives, and IV fluids are suggested. Intensive care with cardiac monitoring may be crucial.

Asters

Machaeranthera species—*Asteraceae,* Sunflower Family

Desert Aster,
Machaeranthera tephrodes

Sticky Aster,
Machaeranthera bigelowi

Asters grow up to 3 feet tall.

Mojave Aster,
Machaeranthera tortifolia

Description

Asters are perennial plants. They have solitary flowers with yellow disks and purple or yellow rays growing at the ends of their woody branches. They accumulate selenium, taking it in from the soil and storing it up in concentration. Animals will generally not eat the plant unless there is no other forage available because it has a strong smell, similar to garlic. However, the plant may be consumed in poor quality hay if animals are very hungry.

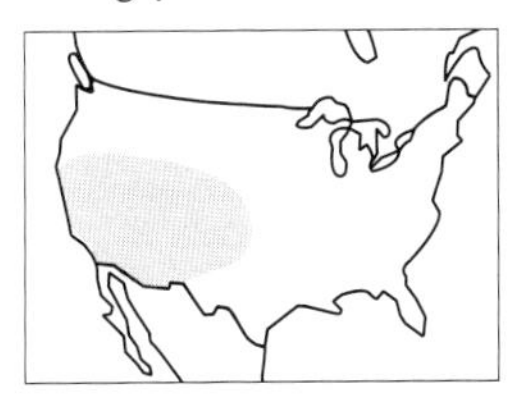

Geographic Distribution

Alkaline soils containing selenium in Colorado, Wyoming, South Dakota, Arizona, and California.

Signs of Poisoning

Animals eating the plant may develop chronic selenium poisoning with loss of long mane and tail hair, horizontal cracking of hooves, lameness, stiff joints, emaciation, rough coat, and staggering gait. Death may occur in acute poisoning as a result of liver and heart disease.

What to Do

Remove the horse promptly from the hay or field. Once signs of chronic selenium poisoning are present, only symptomatic treatment is possible. Meticulous hoof care will help prevent loss of the hoof wall. In selenium rich soils of North America, it is important to feed a balanced ration with adequate sulfur and copper levels to counteract the high selenium in the plants. Alfalfa with high levels of sulfur-containing amino acids is an ideal food source. Dried molasses and bran are good copper, sulfur, and trace mineral sources, but the extra phosphorus in bran must be carefully balanced with a calculated amount of calcium.

Broomweed

Gutierrezia microcephala—*Asteraceae*, Sunflower Family

Broomweed
Up to 2 feet tall

Broomweed

Similar Species

Resinweed, Perennial Snakeweed, Slinkweed, Sticky Snakeweed, Turpentineweed, *Gutierrezia* species

Description

Broomweeds reach about 2 feet in height with dense slender branchlets growing off a woody stem. Alternating leaves are 3/4–2 inches long, and tiny yellow flowers grow at the ends from June to October. It will be eaten in limited amounts by sheep and horses in times of stress and has proven harmful.

Saponins are the poisonous principle found in the leaves. Sticky snakeweed, also containing saponin, is extremely poisonous to sheep and goats but somewhat less to other animals. It is very similar to broomweed, but with a shiny leaf surface and branches that grow up to 21/2 inches long. Horses rarely eat broomweed.

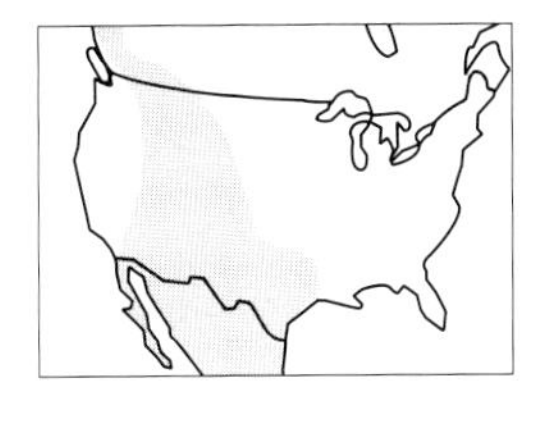

Geographic Distribution

Higher elevations (above 2,800 feet) in the western United States and Canada; on dry range and deserts from the Southwest (particularly Texas) into Mexico, and north into Colorado and Idaho.

Signs of Poisoning

Symptoms of poisoning in livestock may include a rough coat, diarrhea or constipation, nasal discharge, peeling muzzle, kidney and liver degeneration, anorexia, and abortion. Horses may have similar problems.

What to Do

No treatment has been noted, but animals whose diet is correctly balanced and supplemented recover.

Cocklebur

Xanthium species—*Asteraceae*, Sunflower Family

Cocklebur
Up to 4 feet tall

Cocklebur fruit

Cocklebur

Description

The cocklebur is a coarse plant growing to 3 or more feet in height. It has alternating leaves and spiny, burlike fruit shaped like a cockleshell and 11/4 inches long.

There are many species of cocklebur, and all are toxic. Animals generally avoid mature plants and the burs, but the plant's young seedlings (the 2-leafed stage and the most toxic period for this plant) are readily consumed. They have proven fatal even when less than 11/2 percent of the animal's body weight was eaten. Animals can quickly consume this amount in poor quality hay, and drying has little effect on the toxicity level. The burs are also highly toxic but rarely eaten except occasionally by pigs.

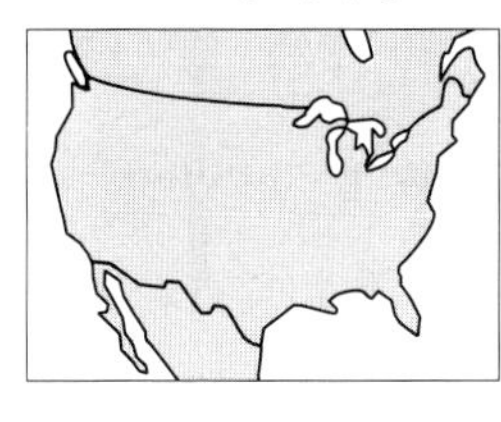

Geographic Distribution

Fields or disturbed soils, near water sources. The plant is hard to eradicate once started due to the seed's persistence before germinating.

Signs of Poisoning

Within a few hours after ingestion, horses will exhibit gastrointestinal distress, depression, anorexia, nausea, weakened heartbeat, muscular weakness, and possibly convulsions. This plant directly affects the liver of animals, and those that survive the acute poisoning frequently develop chronic liver disease.

What to Do

The preferred treatment is orally administered mineral oil and activated charcoal along with other symptomatic treatments, which may include IV fluids and 24-hour intensive care, followed by specialized diets if there is chronic liver damage.

Groundsel

Senecio **species**—*Asteraceae,* Sunflower Family

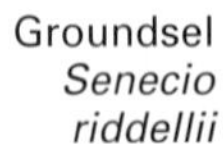

Groundsel
Senecio riddellii

Groundsel
Senecio riddellii

Threadleaf Groundsel,
Senecio longilobus

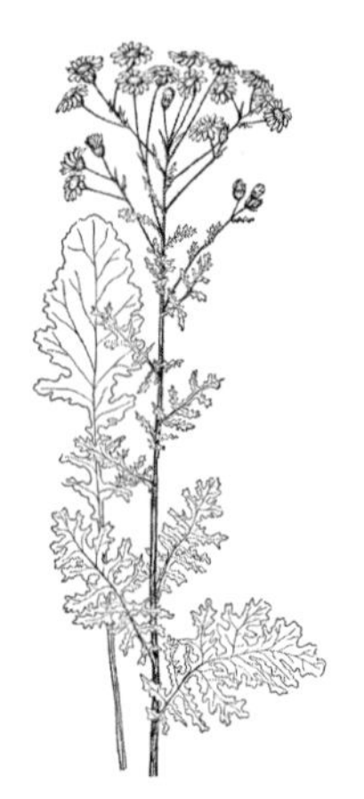

Groundsel
Up to 3 feet tall

Description

Groundsels are perennial herbs growing to about 3 feet tall with branches shooting upright from a woody base. Leaves range in size and form from large and deeply lobed to narrow and elongated. Most have yellow daisylike flowers clustered at the tops of their stems or branches. The distinguishing feature of *Senecios* is the single layer of bracts surrounding the yellow ray flowers.

Pyrrolizidine alkaloids are the toxic principle, and all parts of the plants are toxic, even when dried. Horses will not usually eat groundsels unless other food is scarce or unless they are incorporated in hay.

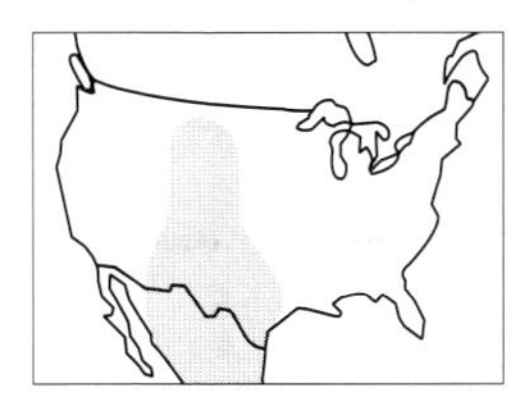

Geographic Distribution

Dry areas in altitudes from 2,500 to 7,500 feet in Utah, Colorado, Arizona, Texas, and Mexico.

Signs of Poisoning

Poisoning may not be evident for 2–8 months after an animal has started eating groundsel. Symptoms in horses and cattle include weight loss, jaundice, depression, abdominal pain, and nervousness. Anemia, frequent urination, and straining to defecate may occur in some animals. Horses may exhibit sluggishness, delirium, or aimless walking—sometimes for miles ("walking disease"). Some horses may show unusual behavior, including aggressiveness. Photosensitivity is likely in white-skinned areas. Animals that have developed clinical signs of liver failure (jaundice, photosensitization) invariably die from severe liver cirrhosis.

What to Do

No treatment is effective once the liver is destroyed by the pyrrolizidine alkaloids. Avoid all *Senecios*, hound's tongue, and tarweed in horse pastures and hay.

Jimmyweed

Haplopappus heterophyllus—*Asteraceae,* Sunflower Family

Goldenbush,
Haplopappus species
Up to 4 feet tall

Goldenbush,
Haplopappus species

Also Known As

Jimmy Goldenweed, Rayless Goldenrod, Burrow Weed

Description

This plant has a woody base and numerous stems stretching upward to approximately 2–4 feet. Tiny yellow flat-topped flowers appear in clumps on the branches from June to September. The fruit is about 1/12 inch long and grows a hairy skin. The leaves are alternate and simple, pale green, growing about 2–3 inches long, and sometimes have hair. The main stem of the plant is gray or white.

It takes only 1 percent of a horse's weight, consumed daily for 3 or more days, to cause poisoning. Foals may die from drinking a mother's contaminated milk, while the mother slowly poisons herself grazing on the plant. The toxic principle is tremetol.

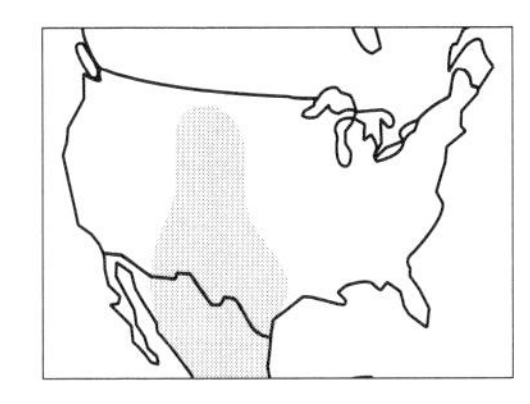

Geographic Distribution

On hillsides in higher altitudes, along waterways, and on roadsides from Mexico into Texas, Arizona, Colorado, and northward.

Signs of Poisoning

After 1–3 weeks, symptoms may include depression, stiffness, constipation, frequent urination, trembling, weakness, shortened pulse, knuckling of the fetlocks and eventual paralysis, and convulsions. Liver and kidney damage will occur after very little consumption of the plant. Horses may develop signs of congestive heart failure.

What to Do

No treatment has been noted. Activated charcoal, saline cathartics given orally, and other supportive treatment is indicated.

Animals not promptly removed from the source will lapse into a coma and die quietly. Pull all jimmyweed in the horse pasture, and do not feed hay containing jimmyweed. Forced exercise may cause the horse to collapse and die.

Snakeroot

Eupatorium rugosum—*Asteraceae,* Sunflower Family

Snakeroot,
Eupatorium rugosum

Snakeroot
Up to 4 feet tall

Also Known As

Richweed, White Sanide

Description

Snakeroot has straight stems growing up to 4 feet tall. The oval or heart-shaped leaves have sharp toothed edges and grow opposite one another. From July to October, white flowers grow in flat-topped clusters at the ends of the stems.

The most dangerous seasons for this plant are summer and fall, but horses and livestock may be poisoned from it at other times of year if the plant gets mixed into their hay.

Tremetol, a complex benzyl alcohol, is the toxic principle found in the leaves and stems. The toxin remains active in the dried plant and is excreted in the milk of lactating mothers that eat the plant in any form.

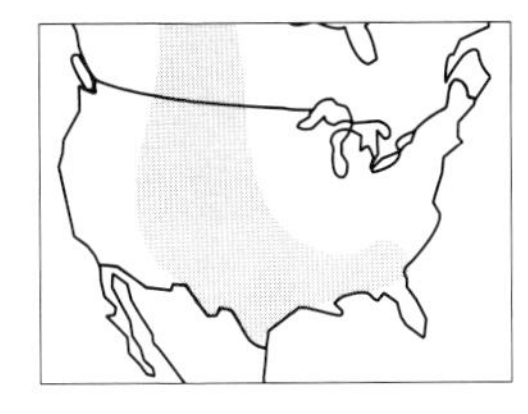

Geographic Distribution

In open woods, along waterways, and in pastures; east of the Rockies from Canada, south to Texas, Louisiana, and Georgia.

Signs of Poisoning

Snakeroot is responsible for "milk sickness" in cattle, which poisons the milk and meat of the animal, and "the trembles" in horses, which causes the animal to have sluggish behavior, slobber, lose its appetite, and become nauseated. More permanent changes from snakeroot poisoning include liver and kidney degeneration and hemorrhages in the organs.

What to Do

Treat symptomatically; activated charcoal and laxatives may be given orally. Severely affected animals may need IV therapy and other intensive care measures.

Sneezeweed

***Helenium* species**—*Asteraceae*, Sunflower Family

Sneezeweed,
Helenium species

Sneezeweed ray-and-disk flowers

Sneezeweed
Up to 3 feet tall

Also Known As

Bitterweed, Staggerwort, Swamp Sunflower

Similar Species

Bitter Sneezeweed, Small Head Sneezeweed, *Helenium* species; Orange Sneezeweed, *Dugaldia hoopesii*

Description

Sneezeweeds are 2- to 3-foot-tall perennials with 1- to 6-inch toothed leaves sparsely alternating along the branches. Their flowers have yellow rays surrounding a darker yellow disk. A smaller variety, *H. nudiflorum*, has flowers with yellow rays surrounding a purple disk.

The poisonous principles are sesquiterpene lactones (helanin), but other toxins may also be present. The mature plant is the most toxic; but all parts, including the seeds, can cause poisoning, and dried plants remain toxic. All livestock and especially sheep are susceptible to sneezeweed poisoning.

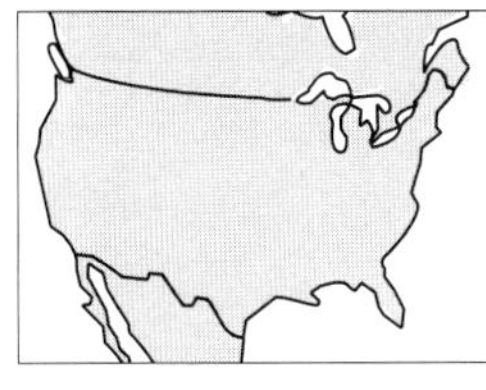

Geographic Distribution

Swamps, moist slopes and meadows throughout most of the United States and Canada. Sneezeweeds become a problem when pastures are overgrazed.

Signs of Poisoning

Weakness, convulsions, foaming at the mouth, and lack of coordination are all symptoms both horses and livestock may exhibit.

What to Do

Intensive care may be needed to help the horse get through the acute poisoning. Activated charcoal and saline cathartics are of benefit if animals have recently eaten the plants. Remove the horse promptly from the hay or field and feed good quality forage.

Yellow Star Thistle

Centaurea solstitialis—*Asteraceae*, Sunflower Family

Russian Knapweed, *Centaurea repens*

Yellow Star Thistle, *Centaurea solstitialis*

Yellow star thistles can grow up to 2 feet tall.

Yellow Star Thistle, *Centaurea melitensis*

Similar Species

Russian Knapweed, *Centaurea repens*

Description

This is an annual plant that grows up to 2 feet tall. Its leaves are covered in cottony hair and it has bright yellow, spiny-based flowers. The slightly taller Russian knapweed has lavender-rose to white thistlelike flowers at the ends of its branching stems.

Poisoning usually occurs in summer and fall, when animals may acquire a taste for the plant and seek it out. Horses seem very susceptible to these plants. Their toxic principle is unknown.

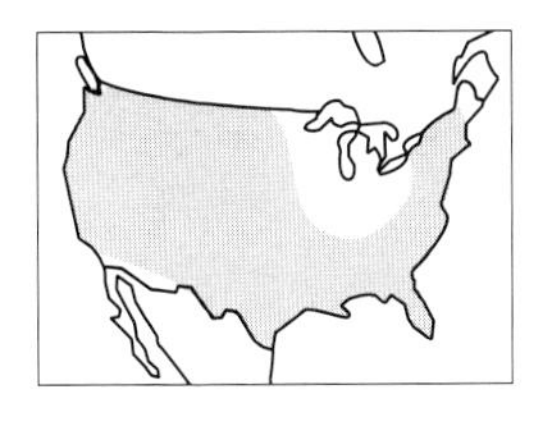

Geographic Distribution

On roadsides and in neglected areas, pastures, and cultivated areas in the eastern, southern, and western United States. Russian knapweed is spreading invasively through the intermountain states.

Signs of Poisoning

"Chewing disease," also known as *nigropallidal encephalomalacia*, may occur in horses if large quantities of the plant are eaten for several weeks or months. At this point, when cumulative effects of the plant reach a toxic threshold, symptoms seem to appear overnight. Involuntary chewing, lip twitching, swelling around the mouth, and poor reflex control prevent the animal from biting and chewing its food. Due to hypertoxicity in the facial muscles, a horse may stand with its mouth agape, unable to eat. Swallowing is not affected, and a horse can drink water if able to submerge its face in a water trough deep enough to get the water to the back of its mouth.

Once clinical signs of poisoning are present, the condition is irreversible and death eventually occurs from starvation, dehydration, and inhalation pneumonia.

What to Do

There is no known effective treatment, and the prognosis is very poor because brain damage is permanent. Oral nutrition via stomach tube will help sustain an affected horse, but complete recovery is unlikely. Euthanasia is often undertaken to spare the horse from starving to death.

Horses should not be kept in areas where they have access to yellow star thistle.

Hound's Tongue

Cynoglossum officinale—*Boraginaceae*, Borage Family

Hound's Tongue, *Cynoglossum officinale* Up to 3 feet tall

Hound's Tongue flowers

Hound's Tongue fruits

Description

Introduced from Eastern Europe, hound's tongue is a biennial that in its first year forms a rosette of hairy tongue-shaped leaves 8–12 inches long. In the second year, small red flowers are produced on a branching stem 2–3 feet high. Fruits consist of 4 flattened nutlets with "Velcro"-like hooks that enable the seeds to stick to clothing and animal fur, thus aiding their dispersion.

The principle toxins are pyrrolizidine alkaloids, with effects identical to those of the *Senecios* (*see* Groundsel). All parts of the plant are toxic in both green and dried states. Hound's tongue is actually more palatable when dried.

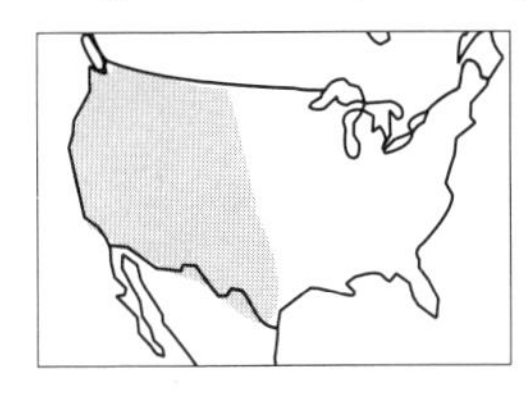

Geographic Distribution

Neglected areas in the western United States.

Signs of Poisoning

The primary toxic effect of pyrrolizidine alkaloids is on the liver, and signs of poisoning often do not appear for several months after an animal starts eating the plants.

Weight loss, photosensitization, depression, and neurological abnormalities are often the first signs observed. Anemia, jaundice, subcutaneous edema, and diarrhea develop as liver failure progresses. Horses exhibiting these signs invariably die from irreversible liver cirrhosis.

What to Do

There is no specific treatment for pyrrolizidine poisoning, and no treatment is effective once the liver is destroyed. Avoid *Senecios* as well as tarweed and hound's tongue in horse pastures and hay.

Tarweed

Amsinckia intermedia—*Boraginaceae,* Borage Family

Tarweed, *Amsinckia intermedia*

Tarweed, *Amsinckia intermedia*

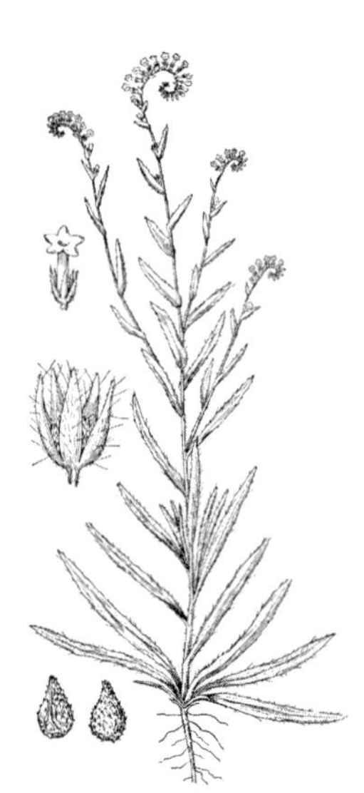

Tarweed
Up to 3 feet tall

Also Known As

Fiddleneck

Description

Tarweed is an annual plant, 1–3 feet in height, with narrow, hairy, pointed leaves alternating along sparsely branched hairy stems. Its small, orange-yellow flowers appear in racemes coiled at the ends of the stems (similar in appearance to a fiddleneck fern), which uncoil as the flowers open. Its seeds are 1- to 2-inch-long, gray to black nutlets.

The principle toxins are pyrrolizidine alkaloids. The seeds of the plant are the most troublesome when they contaminate wheat chaff or grain screenings.

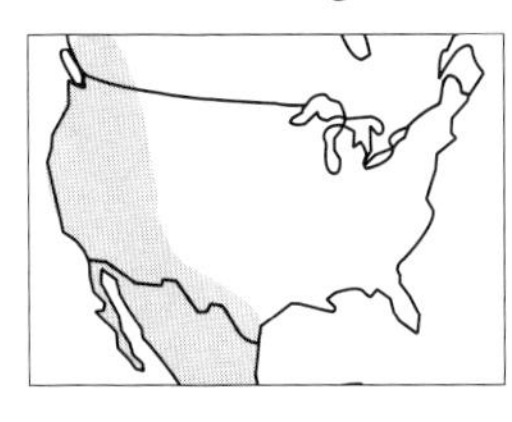

Geographic Distribution

Neglected areas or dry, open cultivated areas (wheat fields) of the Northwest and Pacific Coast states, east into Idaho and south to Arizona.

Signs of Poisoning

Poisoning may not be evident for 2–8 months after an animal has started eating tarweed. Symptoms in horses and cattle include weight loss, jaundice (yellow eyes and mucous membranes), depression, abdominal pain, and nervousness. Anemia, frequent urination, and straining to defecate may occur in some animals. Horses may exhibit sluggishness, delirium, or aimless walking—sometimes for miles ("walking disease"). Some horses may show unusual behavior, including aggressiveness; this is caused by high blood ammonia levels from the liver failure affecting the brain. Photosensitivity is likely in white-skinned areas. Animals that have developed clinical signs of liver failure (jaundice, photosensitization) invariably die from severe liver cirrhosis.

What to Do

No treatment is effective once the liver is destroyed by the pyrrolizidine alkaloids. Avoid all *Senecios*, hound's tongue, and tarweed in horse pastures and hay.

Prince's Plume

Stanleya pinnata—*Brassicaceae,* Mustard family

Prince's Plume,
Stanleya pinnata
Up to 5 feet tall

Prince's Plume
flower spike

Also Known As

Desert Prince's Plume

Description

Prince's plume grows to 1–5 feet tall. It is a perennial with a woody base and simple, thick, coarse stems with 5- to 8-inch-long, pale green, smooth, often-lobed leaves. Branches are green on the young plant and brown when older. The plant blooms from May to June with numerous 3/4-inch yellow flowers on a tall spike. Its narrow seedpods are 2–3 inches long and curve downward.

Prince's plume accumulates selenium and has caused selenium poisoning in some animals that graze on it. It is, however, rarely eaten and is of greatest significance in that it grows only where selenium is present in the soil, thereby acting as a selenium indicator.

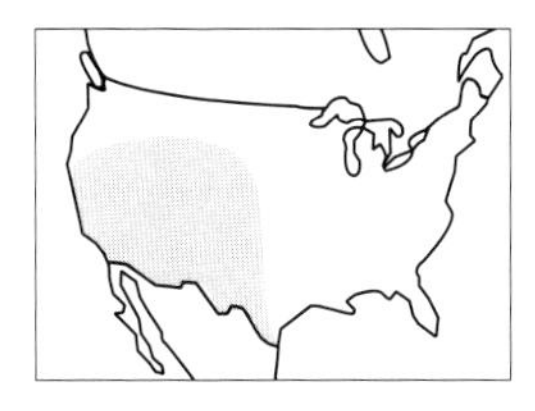

Geographic Distribution

Texas, north through South Dakota and west to California; on dry plains and slopes of alkaline, shale-like soils, and rock formations.

Signs of Poisoning

Animals eating the plant may develop chronic selenium poisoning with loss of long mane and tail hair, horizontal cracking of hooves, lameness, stiff joints, emaciation, rough coat, and staggering gait. Death may occur in acute poisoning as a result of liver and heart disease.

What to Do

Once signs of chronic selenium poisoning are present, only symptomatic treatment is possible. Meticulous hoof care will help prevent loss of the hoof wall.

Remove the horse promptly from the hay or field. In selenium rich soils of North America, it is important to feed a balanced ration with adequate sulfur and copper levels to counteract the high selenium in the plants. Alfalfa with high levels of sulfur-containing amino acids is an ideal food source. Dried molasses and bran are good sources for copper, sulfur, and trace minerals, but the extra phosphorus in bran must be carefully balanced with a calculated amount of calcium.

Rape

Brassica napus—*Brassicaceae,* Mustard family

Kale, *Brassica oleracea*

Wild Radish, *Raphanus raphanistrum*

Wild Radish, *Raphanus raphanistrum*

Rape
Up to 3 feet tall

Weeds & Wildflowers

Similar Species

Canola; Black, Indian, Tansy, White, Wild, and Wormseed Mustards; Charlock; Rutabaga; Kale; Cabbage; Turnips; Horseradish; Wild Radish; Wintercress; Fanweed; Field Pennycress

Description

Considerable variation occurs between the genera, but most members of the family usually have yellow or white flowers with four petals, four stamens, and four anthers.

Most poisoning in animals that eat quantities of the mustard family is due to the glucosinolates present in the plants. These compounds can be converted to various toxic compounds in the animal's body that can cause gastroenteritis, anemia, thyroid enlargement, and kidney and liver disease. Blindness, abortion, and photosensitivity may also result from the actions of the glucosinolates.

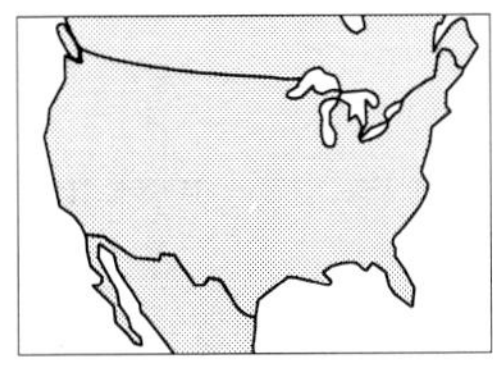

Geographic Distribution

Throughout North America; wild throughout the eastern and midwestern United States; cultivated for use as a condiment. Rape is frequently found in northern pastures.

Signs of Poisoning

Depending on the quantity of plant, seed, or oils (mustard oil) that an animal consumes, symptoms may include colic, diarrhea, jaundice, anemia, weight loss, and thyroid enlargement.

What to Do

Recovery can be expected if the animal is removed from the source of the plants or seeds. Mineral oil via stomach tube will help evacuate the intestinal tract. All sources of the plants, seeds, etc., should be removed from the animal's diet.

Corn Cockle

Agrostemma githago—*Caryophyllaceae,* Pink Family

Corn Cockle,
Agrostemma githago

Corn Cockle
Up to 3 feet tall

Description

Corn cockle is green to grayish and grows up to 3 feet tall; its opposite leaves are 2 to 4 inches long and are coated with little white hairs. From June to September, 2-inch-wide pink or purple flowers with 5 petals grow about 1 inch long at the ends of long slender stems.

The fruit capsules are noted as having small black seeds, and these contain the toxic sapogenin, githagenin. Most poisonings occur when the seeds are mixed with grain for horses and livestock.

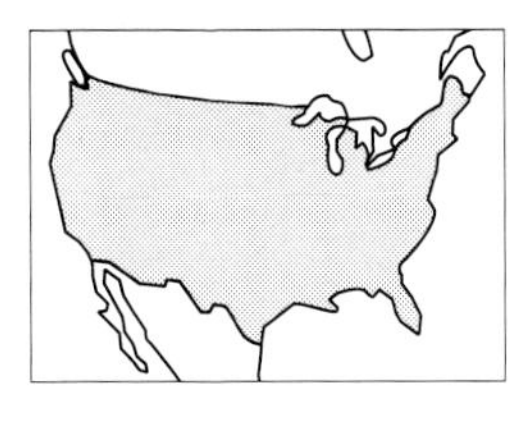

Geographic Distribution

Throughout the United States; along roadsides and in crop areas of winter wheat and occasionally winter rye.

Signs of Poisoning

Poisoning causes gastrointestinal upset, diarrhea, and death.

What to Do

Mineral oil, activated charcoal, and saline laxatives given via stomach tube will help remove the seeds from the intestinal tract. Intensive care and other supportive therapy should be given as needed.

Leafy Spurge

Euphorbia esula—*Euphorbiaceae,* Spurge Family

Field of Leafy Spurge, *Euphorbia esula*

Leafy Spurge flower heads, *Euphorbia* species

Leafy Spurge
Up to 3 feet tall

Similar Species

Snow on the Mountain, *Euphorbia marginata;* Poinsettia, *Euphorbia pulcherrima*

Description

There are more than 1,000 species in this group of plants, all producing a milky sap. Common spurges are annual shrubs with simple, opposite, or alternating leaves. The flowers are unusual, with 4 or 5 yellowish petals surrounded by a cluster of male flowers with single stamens. The center of the female flower eventually ripens into a seed capsule.

Poinsettias are mildly toxic, with new hybrids being even less so. Leafy spurge is very invasive and in particular has caused trouble for horses. Fortunately, however, horses rarely eat spurges unless they are starving.

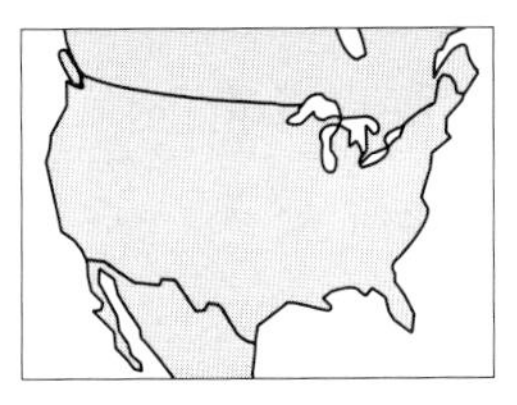

Geographic Distribution

Throughout North America.

Signs of Poisoning

Euphorbias cause blistering of the lips, tongue, and skin; irritation of the gastrointestinal tract; and photosensitization. Freshly mowed spurges can cause inflammation and hair loss on the feet of horses due to the irritant properties of the sap.

What to Do

Removal of the horse from the plant usually brings about full recovery. A demulcent to protect the gastrointestinal tract may need to be administered. Washing the legs with mild soap will help remove any sap that may be present.

Clover, Alsike

Trifolium hybridum—*Fabaceae,* Pea Family

Alsike Clover,
Trifolium hybridum
Up to 2 feet tall

Description

Each leaf on these plants has 3 small, tapered leaflets that branch from their stems. Flowers are creamy white to pink and bloom from May to October. Plants grow from 1–2 feet in height.

The toxic principle is an unidentified phototoxin.

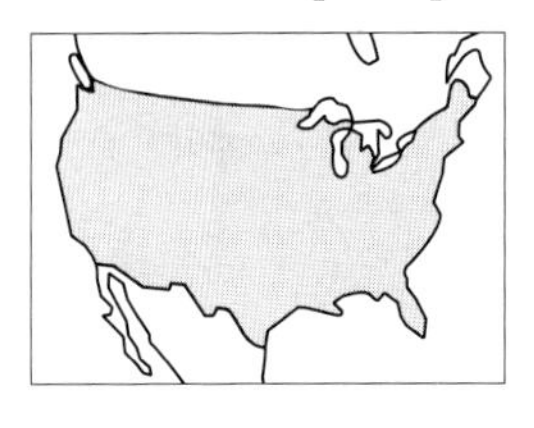

Geographic Distribution

Cultivated; wild in fields and roadsides throughout the United States

Signs of Poisoning

The English were the first to discover that alsike clover caused photosensitivity in horses, now called trifoliosis. A horse contracting trifoliosis will have reddened skin after being turned out in the sun, especially around the muzzle area, tongue, and feet. Lesions on the skin may develop and become severely inflamed. Known as "dew poisoning," the photosensitivity appears to occur when pastures of clover are wet and the horse's skin is moist. The problem could be so acute that the horse may not be able to eat.

Horses also show signs of colic, depression, or excitement, and have diarrhea after consumption. The liver is often affected and enlarged, causing the horse to lose weight and develop jaundice. If the animal continues eating infected pasture or affected hay, it will have recurring bouts of unsteadiness and depression with neurologic disturbances typical of terminal liver disease.

What to Do

Animals removed from the source early have a good chance of full recovery. No other treatment has been noted.

Clover, Crimson

Trifolium incarnatum—*Fabaceae*, Pea Family

Crimson Clover, *Trifolium incarnatum* Up to 6–30 inches tall

Crimson clover, *Trifolium incarnatum*

Description

These plants grow from 6–30 inches in height and have leaves with three blunt-end leaflets. The longish, tapering flower heads are dark red and bloom May–July. Short, stiff, barbed hairs grow beneath the flower heads; these tend to be dangerous to horses when the plant is put into hay at an overripe stage. This plant is frequently used as a forage crop for livestock.

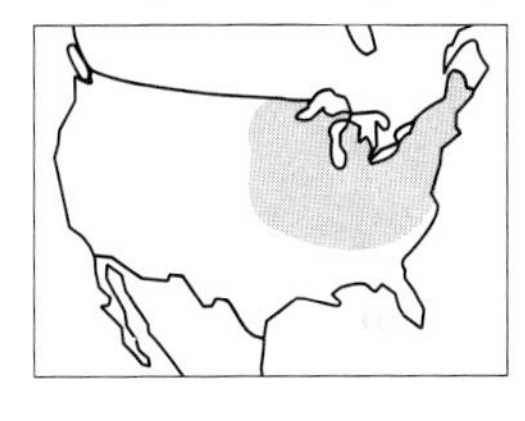

Geographic Distribution

Cultivated; wild in the central eastern United States.

Signs of Poisoning

The ingested material causes colic and impaction in the intestines, which can result in death. Photosensitivity has been known to develop in horses.

What to Do

Surgery may be necessary to remove the impacted fibrous mass of plant material from the intestinal tract if conservative treatment with mineral oil and laxatives is ineffective in moving the impaction.

Clover, Red

Trifolium pratense—*Fabaceae,* Pea Family

Red Clover,
Trifolium pratense

Red Clover

Red clover
Up to 16 inches tall

Description

Red clover has a purple-red, round-headed flower on a hairy stem, and its leaves are divided into three oval leaflets showing pale chevrons on their surfaces. It grows to a height of 6–16 inches and flowers from May to September. If the clover in hay becomes damp, causing it to become moldy (brown spots on the leaves), a fungal toxin, slaframine, is produced, which induces severe salivation, or slobbering.

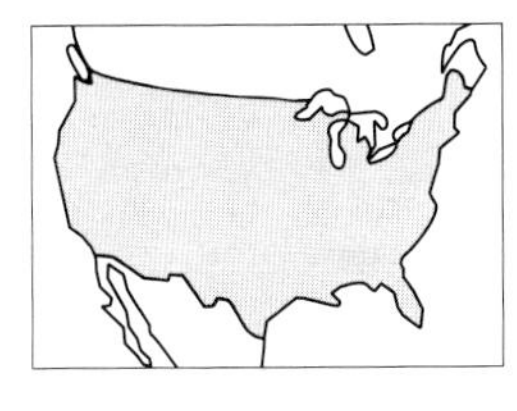

Geographic Distribution

Wild in hay fields, lawns, and along roadsides; throughout the United States.

Signs of Poisoning

Red clover poisoning causes salivation within 30 minutes of ingestion. Bloating, stiffness, diarrhea, blindness, and abortion may occur. Horses may founder if allowed to eat on lush pasture land where the red clover thrives.

What to Do

Remove the horse from the source of the red clover. Antihistamines may be of some benefit if given early. Rehydration of the animal may be necessary where severe salivation has occurred.

Clover, Sweet

Melilotus species—Fabaceae, Pea Family

White Sweet Clover, *Melilotus* species

White Sweet Clover flower spikes

Sweet clover grows up to 5 feet tall.

Description

Sweet clover has erect stems and leaves divided into 3 segments. It has spikes of flowers, white or yellow, that give off a pleasant fragrance when crushed. The white variety can grow as tall as 8 feet, but the yellow variety is less robust.

In itself, sweet clover is not poisonous, but when cut for hay and allowed to become moldy under conditions of high humidity, various fungi metabolize coumarin in the plant to form the toxic principle dicumarol. Poisoning is most likely to occur in cattle, as horses will rarely eat moldy sweet clover hay.

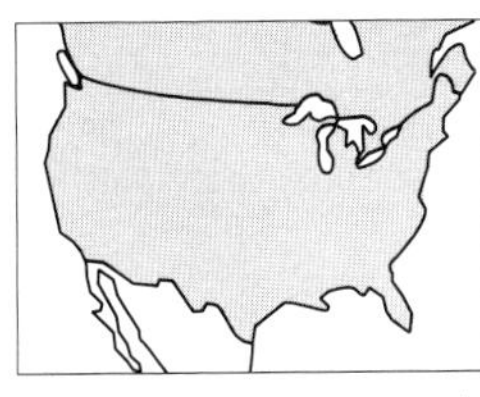

Geographic Distribution

Moist soils throughout the United States and Canada. In some areas sweet clover is grown for hay and compares favorably with alfalfa in nutrient value. In other areas, such as in Texas, sweet clover has become a weed invading pastures, roadsides, etc.

Signs of Poisoning

Horses and livestock show symptoms within 3–8 weeks after ingesting moldy sweet clover. Symptoms include anemia, weakness, abnormal heartbeat, lameness, and abortion. Swellings may appear over bony protuberances of the body due to bruising and hematoma formation. Dicumarol interferes with normal blood clotting and results in hemorrhaging; if this occurs in the brain or other organs, the horse will likely die within 3 days. Minor surgical procedures can hemorrhage profusely.

What to Do

Massive blood transfusions and vitamin K may help. Remove all moldy sweet clover hay from the animal. Where severe blood loss has occurred, blood transfusions are necessary to provide clotting factors. Animals should be handled with care to avoid stress and further hemorrhaging. Vitamin K1 has been shown to be effective in reversing the clotting effect. Vitamin K3 should be used very cautiously in horses, if at all.

Coffeeweed

Cassia species—*Fabaceae,* Pea Family

Partridge Pea, *Cassia* species

Partridge Pea, *Cassia fasiculata*

Most sennas grow up to 3–5 feet tall.

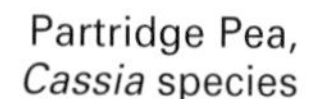

Partridge Pea, *Cassia* species

Also Known As

Coffee Senna, Wild Coffee

Similar Species

Senna, Partridge Pea, *Cassia fasiculata;* Coffeepod, Sicklepod, *Cassia obtusifolia*

Description

Most sennas are erect annuals or perennials about 3–5 feet tall, with yellow or orange flowers that wilt in the heat of the afternoon sun. The gray or green pinnately compound leaves are alternate and some are spirally arranged; they have an unpleasant odor. Seed pods are flat and brown with cross partitions, and seeds are the the plant's most toxic part.

Although all the toxins are not yet known, N-methylmorpholin and an anthraquinone glycoside have been found in senna plants. The toxin has both a cathartic effect and causes myodegeneration of skeletal and cardiac muscle. Liver degeneration may also occur.

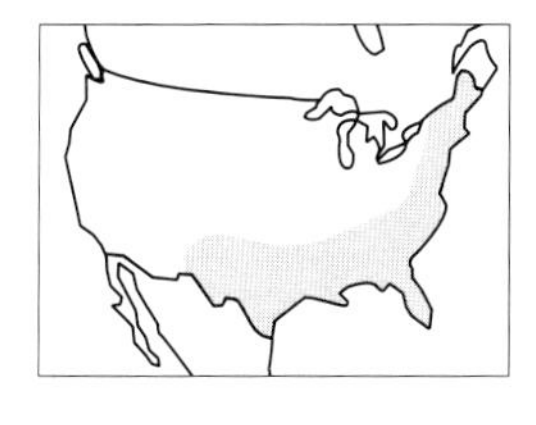

Geographic Distribution

Moist open woods, neglected or disturbed areas, pastures, and roadsides; throughout eastern and southern United States.

Signs of Poisoning

The poison is cumulative, and it takes several days to weeks for symptoms to appear. In horses, these signs include diarrhea, anorexia, staggering gait, and brown or red urine. Muscle degeneration will cause horses to fall and be unable to rise; degeneration of the liver and kidneys is also present, and death will result from degeneration of the heart muscle. The pale muscles observed during post-mortem exams are the reason the affliction has been called "White Muscle Disease"—not to be confused with selenium deficiency. Throughout the poisoning, horses remain alert.

What to Do

Animals may recover if removed from access to the plant and if muscle, liver, and kidney degeneration is not severe. Intravenous fluids given to maintain renal function will aid in recovery. There is no specific antidote to the toxin.

Locoweed

Astragalus species—*Fabaceae,* Pea Family

Milk Vetch, *Astragalus bisulcatus*

Milk Vetch pods, *Astragalus bisulcatus*

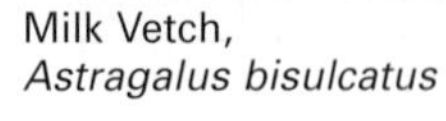

Milk Vetch, *Astragalus bisulcatus*

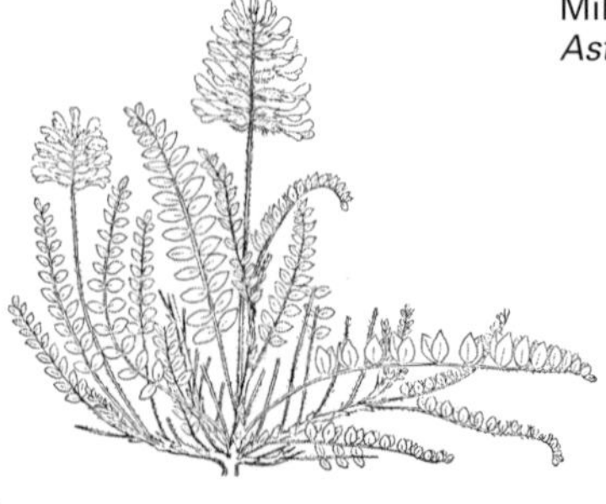

Locoweed
Up to 18 inches tall

Similar Species

Milk Vetch, *Astragalus* species

Description

Locoweeds are low-growing perennials about 18 inches tall. Their alternately arranged, pinnately compound leaves are about 6 inches long and have up to 12 pairs of narrow, woolly leaflets. It has densely packed, hairy spikes of lilac to purple flowers; the two bottom petals join to form a blunt end; some species in the genus have greenish white or creamy blossoms. The fruit pods contain kidney-shaped seeds.

Most *Astragalus* species are poisonous and have caused extensive loss of animal life in the western United States. The poisonous principle is the alkaloid Swainsonine, which causes complex sugars to accumulate in the brain and other tissues, thereby impairing cellular function. Some species of *Astragalus* accumulate selenium while others contain nitroamines that affect the nervous system. Horses find locoweeds palatable and will eat them at any time. When turned out, they may head directly for the locoweed patch! Horses can also be poisoned from eating the dried plants in winter.

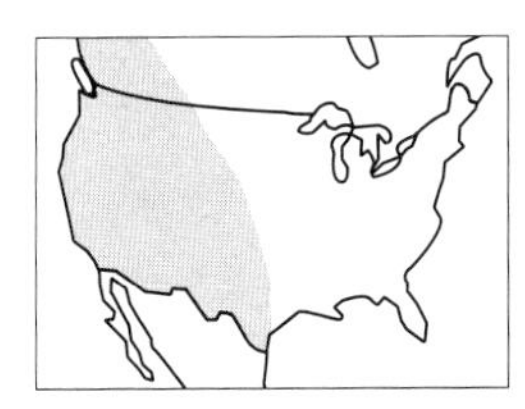

Geographic Distribution

Southwestern Canada, south to Minnesota, South Dakota, Texas, and beyond; prairies and sandy soil, especially in arid and mountainous areas of the West.

Signs of Poisoning

Poisoning usually occurs in late winter or early spring, and symptoms may take several months to appear. Weight loss and abnormal behavior are typical of locoweed poisoning. Horses develop unpredictable behavior, including falling over and charging through fences. Affected horses should not be ridden.

What to Do

There is no effective treatment for locoweed poisoning and death will occur if the animal is not removed from the source. But once clinical signs of locoism are apparent, horses rarely recover completely even if removed from the locoweeds. A "locoed" horse should be given a poor prognosis as there will always be a risk to someone riding that horse. Pregnant mares eating woolly locoweed may produce foals with crooked legs. Prevention of locoweed consumption may be accomplished by holding animals out of locoweed areas until other forages are abundant, and by selective use of appropriate herbicides.

Locoweed, Point

Oxytropis species—*Fabaceae,* Pea Family

Point Locoweed, *Oxytropis viscida*

Blackish oxytrope, *Oxytropis nigrescens*

Point Locoweed

Point Locoweed
Up to 18 inches tall

Weeds & Wildflowers

Similar Species

Point Vetches

Description

Oxytropis and *Astragalus* plants form the largest group of the legume (or pea) family, and individual species are difficult to identify. (Please refer to the Locoweed description.) The sweet-scented flowers of *Oxytropis* species may be distinguished by their lowermost petals, which are prolonged into a point, hence "point locoweed." Its leaves emerge from the ground without stems from a deep root.

Like some of the *Astragalus* species, *Oxytropis* species contain the alkaloid Swainsonine, which causes complex sugars to accumulate in the brain and other tissues, thereby impairing cellular function.

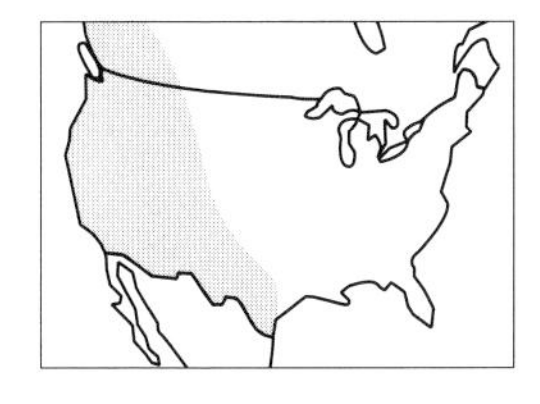

Geographic Distribution

Arid prairies and mountainous area of southwest Canada and western United States.

Signs of Poisoning

Signs of poisoning may take several months to appear, but the poisoning usually occurs in late winter or early spring. Weight loss and abnormal behavior are typical of locoweed poisoning. Horses typically develop unpredictable behavior, including falling over, and charging through fences. Affected horses should not be ridden.

What to Do

There is no effective treatment for locoweed poisoning, and death will occur if the animal is not removed from the source. But once clinical signs of locoism are apparent, horses rarely recover completely even if removed from the locoweeds. A "locoed" horse should be given a poor prognosis as there will always be a risk to someone riding that horse.

Prevention of locoweed consumption may be accomplished by holding animals out of locoweed areas until other forages are abundant, and by selective use of appropriate herbicides.
(*See* Locoweed.)

Lupine

Lupinus species—*Fabaceae,* Pea Family

Lupine, *Lupinus* species Up to 3 feet tall

Lupine flower raceme

Lupine, *Lupinus* species

Lupine

Also Known As

Bluebonnet

Description

These annual plants have basal rosettes of alternate, palmately compound leaves. In spring, racemes of pealike flowers are produced at the ends of the stems; these may be white, red, blue, or yellow.

Quinolizidine and piperidine alkaloids as well as nitrogen oxides have been found in lupines. But not all of the roughly 100 lupine species in North America are poisonous. Toxicity varies with the species and growing conditions (the Texas bluebonnet is believed to be low in toxicity); but if a lupine is toxic it remains so even when dried in hay. The seeds are the most poisonous part of these plants. Sheep are most commonly poisoned by lupine, but horses can also be severely affected.

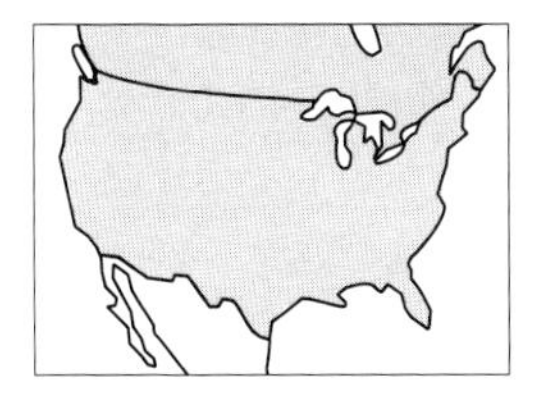

Geographic Distribution

Throughout the United States and Canada; moist to arid soils along roadways and in fields, open woods, and mountainous areas.

Signs of Poisoning

Symptoms will be seen within 1 to 24 hours after ingestion. They include weakened pulse and respiration as well as depression, nervousness, convulsions, and then coma preceding death. Animals that survive may have liver damage and thrive poorly.

What to Do

There is no specific treatment except intensive supportive care. Horses with extensive liver damage are unlikely to recover.

Rattlebox

Crotalaria species—*Fabaceae,* Pea Family

Rattlebox

Showy Crotalaria,
Crotalaria spectabilis

Small Rattlebox,
Crotalaria pumila

Rattlebox
Up to 4 feet tall

Also Known As

Crotalaria

Description

The rattlebox grows to about 2–4 feet tall and has whitish hairs covering its entirety. Its simple, hairy leaves are 2 inches long, with those at the base being oval in shape, while top leaves are more pointed. It has racemes of small yellow flowers that bloom in August–October and black kidney-shaped pods containing seeds that rattle.

The known toxic principle is pyrrolizidine alkaloid (monocrotaline), which induces liver disease.

Geographic Distribution

Eastern and central United States; fields and roadsides, with heavy concentrations in rich bottom lands. This plant has caused many fatalities in horses that are pastured in such areas.

Signs of Poisoning

Symptoms occur within a short period of time after consumption. They include weakness, diarrhea, stupor alternating with improvement, walking in circles; death may occur within a few weeks or months.

What to Do

No treatment has been noted. Hay containing rattlebox has proved to be just as toxic as the live plant and should be disposed of properly.

Vetch, Hairy

Vicia villosa—*Fabaceae,* Pea Family

Hairy Vetch, *Vicia villosa*

Hairy Vetch
flower raceme

Winter Vetch,
Vicia villosa

Vetch
Up to 3 feet tall

Also Known As

Vetch, Winter Vetch

Similar Species

Fava Bean

Description

Hairy vetch grows up to 3 feet tall and has pinnately compound leaves with inch-long, narrow leaflets and curling tendrils. Its tubular flowers may be violet, rose, or white, and they are carried on long stems. These plants have fruit pods.

The toxins of hairy vetch have been poorly defined, but horses have had reactions to the plant. Other vetches are similar in appearance and may also cause problems.

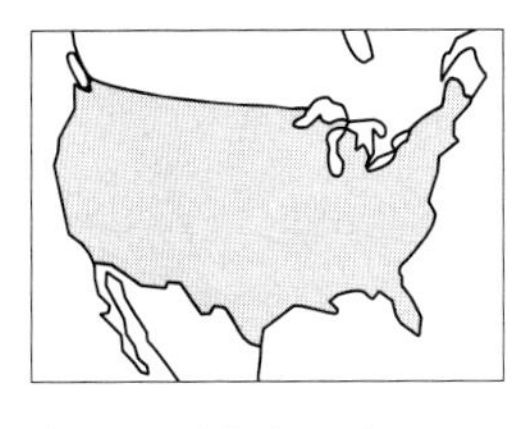

Geographic Distribution

Cultivated; wild in fields and thickets; throughout the United States and Europe.

Signs of Poisoning

Varied symptoms of poisoning include excitement, slobbering, nasal discharge, cough, stiffness, anorexia, weakness, convulsions, swellings of the head and neck, loss of hair, rough coat, diarrhea, and abortion. Liver and kidney damage may be apparent. Photosensitivity is said to occur in pinto horses.

What to Do

No treatment has been noted.

Fitweed

Corydalis species—*Fumariaceae* Family

Corydalis,
Corydalis caseana
Up to 3 feet tall

Golden corydalis,
Corydalis aurea

Corydalis,
Corydalis micrantha

Similar Species

Corydalis, Fumatory

Description

Fitweed is a gray, bushy succulent that grows to 3 feet tall and has cream or yellow spurred flowers, sometimes with purple tips. The flowers grow at the ends of the branches.

Much like the poppy, they contain toxic isoquinoline-structured alkaloids. The plant is toxic at all stages of growth and all species are potentially poisonous.

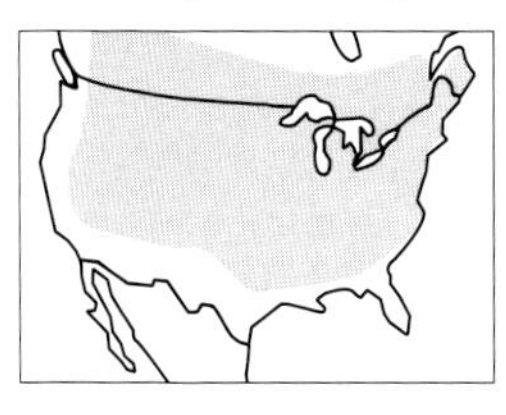

Geographic Distribution

Corydalis caseana is found in the Sierra Nevada Mountains, at elevations of 5,000–6,300 feet and near water. Other *Corydalis* species occur elsewhere in North America.

Signs of Poisoning

Animals eat the plant readily and may die within only a few hours. Panting, seizures, staggering, and snapping with the mouth at nearby objects are all commonly noted within minutes of ingestion of the plant. Depression, twitching (especially in the facial area), and convulsions may also occur.

What to Do

No treatment has been noted, but animals not lethally poisoned recover rapidly once removed from the source of the plant.

St. Johnswort

Hypericum perforatum—*Hypericaceae,* St. Johnswort Family

Common St. Johnswort, *Hypericum perforatum*

Common St. Johnswort, *Hypericum perforatum*

St. Johnswort
Up to 3 or 5 feet tall

Also known As

Klamathweed, Goatweed, Tipton Weed

Description

These perennials grow from 1 to 3 feet high (as tall as 5 feet in Pacific Coast states), with clustered woody stems. The leathery leaves are 1/2- to 1-inch long, grow opposite one another, and have tiny black dots that look translucent when held to the light. Bright yellow, 1-inch flowers with prominent stamens grow in clumps and may be seen from June through September. The fruit is a three-chambered capsule about 1/4 inch long.

At least six of the 25 species of *Hypericum* cause poisoning in animals, and horses and livestock can be affected by the plant whether it is dry or green. New growth is as toxic as the mature plants and is most attractive to grazing animals. The toxic principle is the plant pigment hypericin.

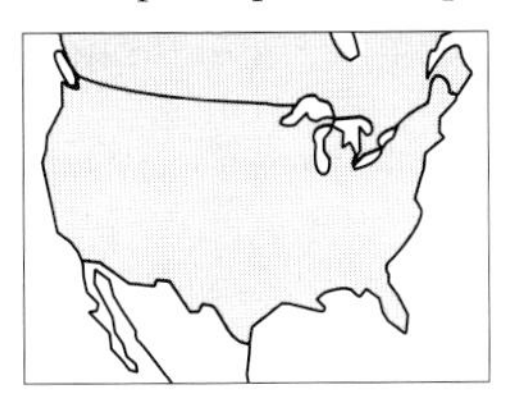

Geographic Distribution

Throughout the United States and Canada; on dry soils along roadsides, in pastures and ranges.

Signs of Poisoning

After ingestion, the toxin is carried via the bloodstream to the skin, where, in the presence of sunlight, it causes photosensitivity in the non-pigmented skin. White-skinned animals are highly susceptible and may rub spots raw or cast themselves to find relief from the intense skin reaction. Elevated temperature, increased heartbeat, diarrhea, and sensitivity to cold water may be seen with St. Johnswort poisoning.

What to Do

Remove the horse from the source of St. Johnswort, and keep the horse out of the sun. Corticosteroids and broad-spectrum antibiotics may help. If the animal is not removed from the sun and the toxic plant, it may suffer from blindness and photosensitivity.

Beefsteak Plant

Perilla frutescens—*Lamiaceae,* Mint Family

Beefsteak Plant,
Perilla frutescens
Up to 3 feet tall

Beefsteak Plant,
Perilla frutescens

Also Known As

Perilla Mint, Purple Mint

Description

Beefsteak is a mint that grows to about 3 feet in height. It has square, branching stems with opposite, toothed leaves up to 5 inches long, with a minty aroma. Smaller leaves surround the single white or purple flowers growing along the stems. The plant's 1/2-inch-long seeds grow in a hairy pod that may be 4 inches long. The underside of the leaves is purple.

Perilla ketone, egomaketone, and isoegomaketone are all responsible for poisoning horses and livestock. These same toxins may also be found in moldy sweet potatoes. Animals generally do not graze on this mint but may consume it in poor quality hay.

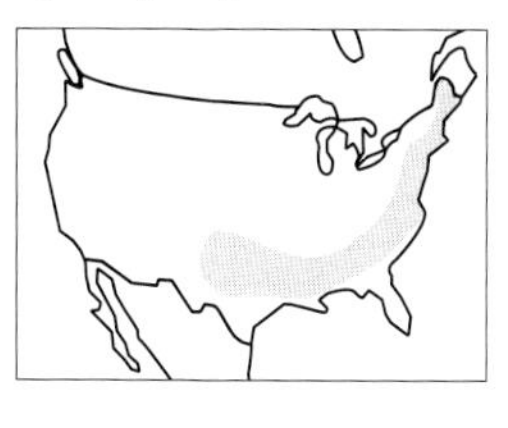

Geographic Distribution

Northern Texas, parts of Oklahoma, Arkansas, and Louisiana, as well as in the eastern United States. Beefsteak originally came from Asia and was introduced to the United States from India.

Signs of Poisoning

Animals develop a severe interstitial pneumonia characterized by sudden onset of acute respiratory difficulty.

What to Do

There is no specific treatment, but diuretics, parenteral steroids, antihistamines, and antibiotics may help if given early. Handle animals gently and prevent stress.

Ground Ivy

Glechoma hederacea—*Lamiaceae,* Mint Family

Ground Ivy,
Glechoma hederacea
Up to 6–8 inches tall

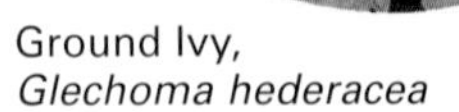

Ground Ivy,
Glechoma hederacea

Ground Ivy,
Glechoma species

Also Known As

Creeping Charlie

Description

Ground ivy is a perennial plant that grows along the ground and has square stems and 1- to 2-inch leaves growing opposite each other. Small blue flowers may be seen in spring and summer. When ground ivy is started, it tends to choke out other vegetation with its complex intertwining stems.

The plant's toxicity comes from a variety of volatile oils it produces. It is considered toxic to horses, but not to livestock. Horses are rarely affected, however, if provided a nutritious diet. And though they may freely graze on the plant, or may eat it in weedy hay, poisoning will occur only if large amounts are ingested.

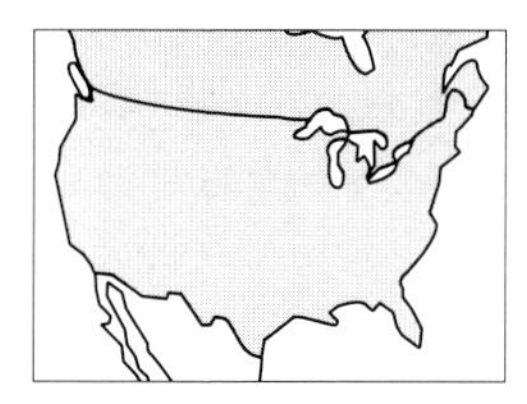

Geographic Distribution

Cultivated as a ground cover; wild throughout the United States and Canada in moist, shaded areas.

Signs of Poisoning

Symptoms after ingestion include salivation, sweating, shortness of breath, and dilated pupils.

What to Do

No treatment has been noted.

Crocus, Autumn

Colchicum autumnale—*Liliaceae,* Lily Family

Autumn Crocus,
Colchicum species
Up to 8–12 inches tall

Autumn Crocus

Autumn Crocus

Weeds & Wildflowers

Description

The plant may be identified by its large grasslike leaves, which grow vertically from the underground bulb in springtime. Clusters of crocus-type white or light purple flowers replace the leaves in the fall.

The entire plant is known to be poisonous to horses and livestock; it contains the alkaloid colchicine as well as other toxic substances. It takes only .1 percent of the animal's weight in leaves to cause poisoning. It is therefore potentially a problem for horses.

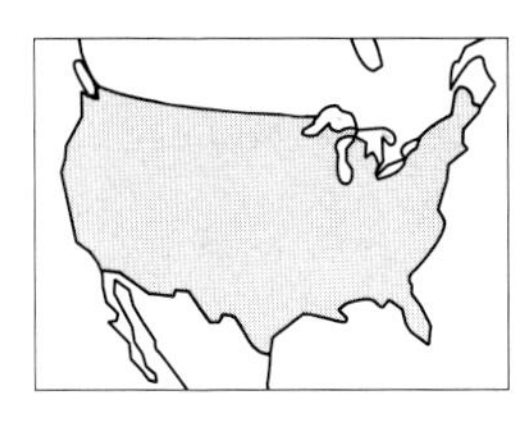

Geographic Distribution

Cultivated as a garden flower throughout the United States; wild in deserted fields.

Signs of Poisoning

Gastrointestinal distress may be the only symptom noted, but grazing on the plant or consuming the bulbs can prove to be fatal.

What to Do

No treatment has been noted.

Death Camas

Zigadenus species—*Liliaceae,* Lily Family

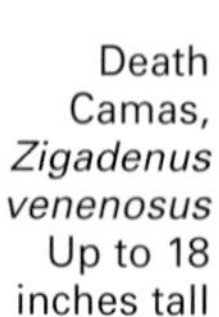

Death Camas, *Zigadenus venenosus* Up to 18 inches tall

Death Camas flower head

Death Camas, *Zigadenus venenosus*

Description

There are approximately 15 species of this plant; all but a few are toxic, and the poisonous varieties are hard to distinguish from the nonpoisonous. They have grasslike leaves and underground bulbs; fruit capsules and greenish yellow or pink flowers are produced in early spring.

The steroid alkaloids from the veratrum group are responsible for extensive poisoning of livestock. The young green plant may be very attractive to hungry animals in the springtime.

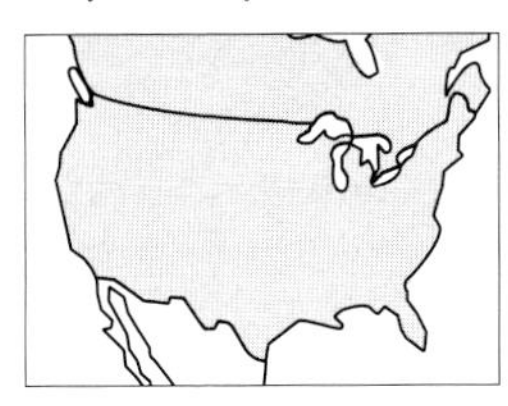

Geographic Distribution

Foothill grazing lands, low-lying woods, and boggy fields, throughout Canada and the United States.

Signs of Poisoning

Depending upon the toxicity level of the plant and the animal that ingests it, the symptoms will vary. Nausea, muscular weakness, weakened heartbeat, convulsions, salivation, and staggering may be noted in horses and livestock, followed by coma and death.

What to Do

Injections of atropine sulfate and picrotoxin may be helpful. Activated charcoal orally may help prevent further toxin absorption.

False Hellebore

Veratrum viride—*Liliaceae,* Lily Family

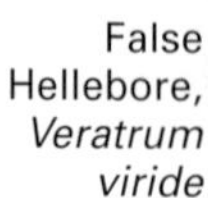

False Hellebore, *Veratrum viride*

False Hellebore bursting through leaf litter in springtime

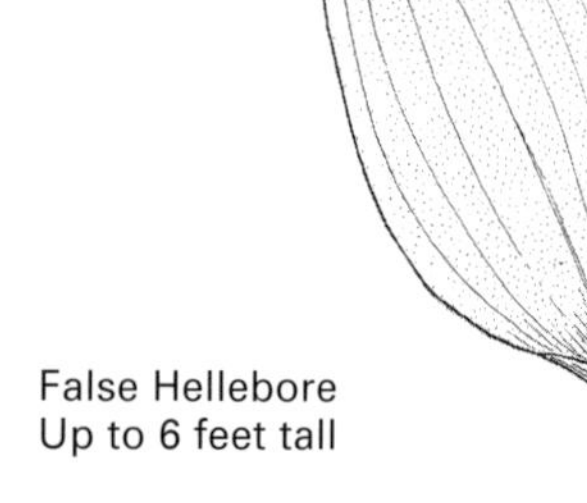

False Hellebore
Up to 6 feet tall

Also Known As

Indian Poke, Skunk Cabbage

Similar Species

Western False Hellebore, *Veratrum californicum*; Corn Lily

Description

False hellebore is a perennial, growing up to 6 feet with unbranched, leafy stems. Oval leaves grow alternately up to 12 inches long and 6 inches wide with parallel veins. Numerous white or greenish flowers are produced on tall flower spikes above the leaves.

The plant contains many steroidal alkaloids and glycoalkaloids. Livestock generally will not eat false hellebore unless other forage is unavailable. Most poisonings occur in spring when the plant is the most attractive to livestock, and also the most dangerous.

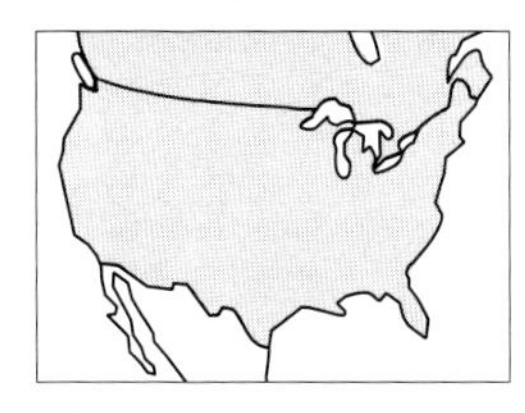

Geographic Distribution

Open fields, low moist woods, and high mountain valleys, throughout the United States and Canada; cultivated in some areas for use as an insecticide and for medical purposes.

Signs of Poisoning

If an animal accidentally ingests false hellebore in its hay or while grazing, within 2 to 3 hours salivation, depressed heart action, digestive upset, diarrhea, respiratory paralysis, and death may be expected. Pregnant livestock and horses that survive are likely to abort their young or give birth to severely malformed fetuses. Deformed offspring usually die at or shortly after birth.

What to Do

There is no specific treatment available. Animals should be prevented from eating the plants especially during pregnancy. Once removed from the plant, recovery usually occurs, unless the fetus is affected in utero. Supportive therapy may be needed in acute poisoning.

Onions

Allium species—*Liliaceae,* Lily Family

Onion, *Allium* species

Blooming Chives, *Allium* species

Onions
Up to 2 feet tall

Similar Species

Garlic, *Allium canadense*

Description

Hollow, rounded grasslike leaves up to 2 feet tall emerge from the underground bulb. Onions can be easily differentiated from grasses on the basis of the onion odor present in all parts of the plant. Flowers are produced as showy terminal clusters, usually white, but various colors are possible.

Onion poisoning most commonly occurs in spring, when young green shoots are developing and are attractive to animals. Livestock as well as dogs and cats also may be poisoned from eating surplus onion bulbs. The toxin in onions is N-propyl disulfide, which is a potent oxidizer of hemoglobin in the red blood cells.

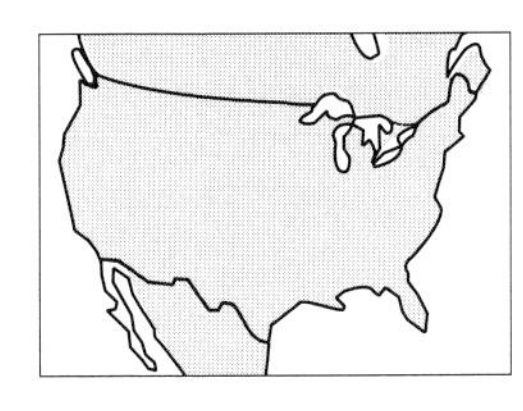

Geographic Distribution

Worldwide; fields and rich wooded areas in the northeastern and north central United States and less frequently in the drier soils of the Southwest.

Signs of Poisoning

Large quantities of the plant may cause gastrointestinal distress, anemia, depression of the red blood cell count, and possibly death, in which case the animal's tissues will have a strong odor of onion. Liver and kidney damage may occur. Many times, symptoms are not recognizable for 1 to 6 days after ingestion of the green plant or onion.

What to Do

If symptoms have just begun, remove the horse promptly, change the diet, and give stall rest. Call your veterinarian to be sure the horse is not suffering from another kind of ailment. Blood transfusions are also recomended if the horse is severely anemic.

Spoiled onions should never be fed to horses, nor should horses be allowed to graze where onions are cultivated or grow wild.

Sacahuista

Nolina texana—*Liliaceae,* Lily Family

Sacahuista,
Nolina species
Up to 5 feet tall

Also Known As

Beargrass

Description

This perennial has long, thin leaves growing densely upward from the bottom of the plant. The main stem grows to 5 feet tall from the center and may have many small stems with clusters of white flowers at their ends as well as a fruit capsule.

The poisonous principle is unknown, but the plant is toxic at all stages of growth, and toxicity is greatest in the flowers. Animals often eat the plant when other forage is unavailable. Toxicity is not lost when the plant is dried, and it may easily be consumed in poor quality hay.

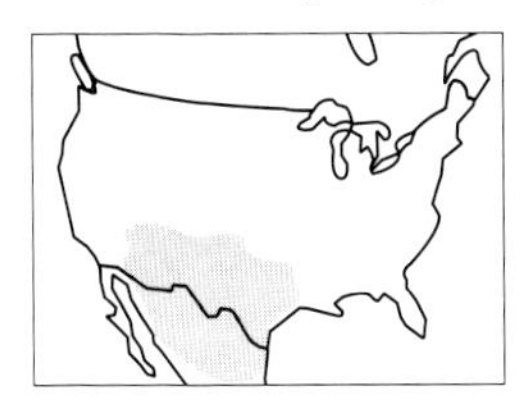

Geographic Distribution

Rangeland in Texas, Arizona, and Mexico.

Signs of Poisoning

Eating the flowers can cause liver and kidney dysfunction (with few other symptoms), leading to death. Eating the buds and fruit (also toxic) may cause photosensitization, anorexia, prostration, dark urine, or yellowish discharge from eyes and nostrils.

What to Do

No treatment has been noted. During the spring blooming season—the most dangerous time for this plant—animals should be removed from areas where the plant occurs.

Mistletoe

Phoradendron villosum—*Loranthaceae*, Mistletoe Family

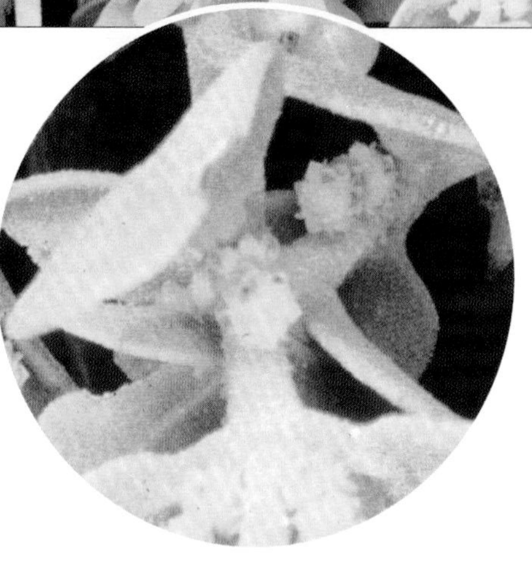

Mistletoe
Up to 4 feet across

Weeds & Wildflowers

Description

The mistletoe is a dense, light-green parasitic plant of 1–4 feet in diameter. It has opposite leaves and pinkish, oval berries. Some varieties have leaves that are as small as the tiny berries. Small flowers—either solitary or in clusters—may be seen from May to July.

Though mistletoe is a favorite plant in the United States around Christmastime, it is toxic at all times of the year. Its poisonous properties may include amines, toxic proteins, and other, unknown elements. Some grazing animals actually like its taste and may seek it out.

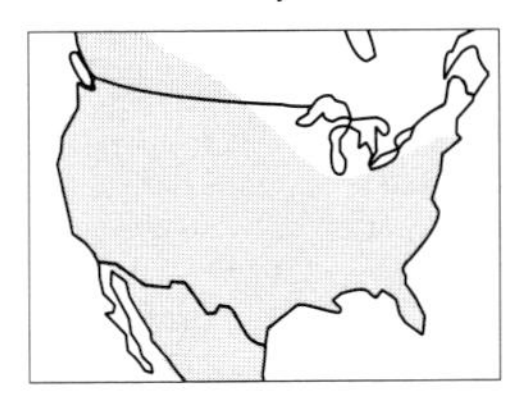

Geographic Distribution

Throughout the United States, mainly in deciduous hardwood trees, most often oaks.

Signs of Poisoning

Horses browsing among oak branches may be poisoned by mistletoe without their owners noticing. After ingestion, few symptoms are seen before a sudden death. Gastrointestinal upset, including colic and diarrhea, may develop in less severe poisoning.

What to Do

There is no specific treatment. Activated charcoal via stomach tube, saline cathartics, and intensive supportive care may be helpful.

Cotton

Gossypium species—*Malvaceae,* Mallow Family

Cotton in bloom, *Gossypium* species

Cotton seed pod

Cotton grows up to 4–5 feet tall.

Cotton

Description

Cottonseed is not a good food for horses. Unfortunately, it is a common component in poor quality horse pellets and sometimes may be mixed into better quality pellets at the mill and not recorded on the feed bag ingredient list.

Cottonseed is high in protein (40 percent) and extremely low in vitamin A and some amino acids. The seeds contain high levels of the toxin gossypol. Processing of the seeds with water or heat should reduce the level of gossypol to a safe level. Ineffective processing can result in toxic levels of gossypol remaining in the cottonseed cakes commonly fed to horses. Safe gossypol levels in the feed should be less than 0.1 percent, and the ration should be supplemented with vitamin A and essential amino acids.

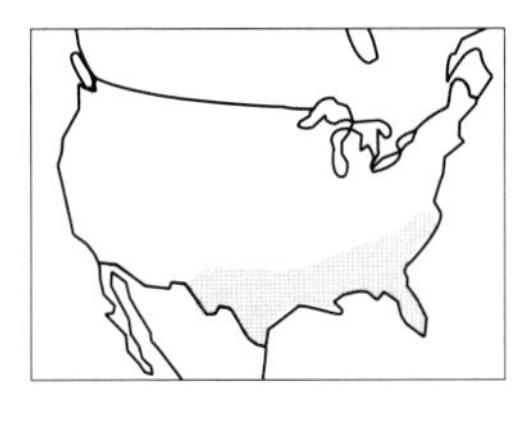

Geographic Distribution

Cultivated in the southern United States.

Signs of Poisoning

Gossypol has a cumulative effect on a horse's system, resulting in poor growth rates, pot belly, gastrointestinal distress, anorexia, joint swelling, congestion, edema, and congestive heart failure.

What to Do

A and B-complex vitamins, high quality protein from varied sources, as well as roughage may help to counteract the effects of toxicity. Treatment of the heart failure is beneficial. Further feeding of cottonseed should be stopped.

Squirrel Corn

Dicentra canadensis—*Papaveraceae,* Poppy Family

Squirrel Corn, *Dicentra canadensis* Up to 1 foot tall

Squirrel Corn flowers

Dutchman's Breeches, *Dicentra cucullaria*

Dicentra species

Also Known As

Staggergrass

Similar Species

Dutchman's Breeches, Bleeding Heart, *Dicentra* species

Description

Dicentras are delicately leaved, early-spring-flowering, herbaceous annuals. The leaves have a lacy appearance similar to a carrot's and originate from the base. The flowers are showy, with four petals fused together to make two pairs, and occur in varying shapes: Squirrel corn's flowers are heart-shaped, while its roots resemble grains of corn; Dutchman's breeches flowers are white and pantaloon-shaped; bleeding heart flowers are red and heart-shaped. The flowers of all these plants droop from leafless stems.

Half of the one dozen species are poisonous to livestock. Horses are rarely affected. The toxin is an isoguinoline alkaloid.

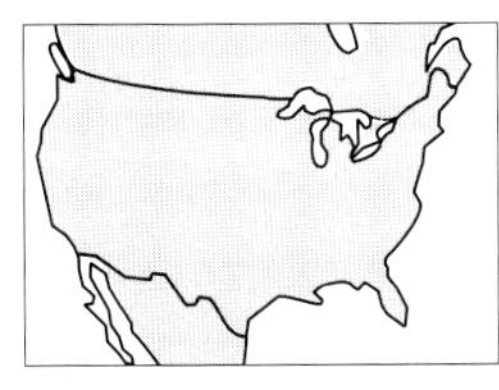

Geographic Distribution

Woodlands, fencerows, and clearings, throughout North America. Squirrel corn is found in the Northeast, south to North Carolina and Tennessee. Dutchman's breeches grows in the Northeast, south to Alabama. Western bleeding heart grows in British Columbia, Oregon, and northern California. The wild bleeding heart is very similar but is found in rocky areas of the southern Appalachians. The common bleeding heart is grown throughout the country as an ornamental plant and is poisonous.

Signs of Poisoning

Symptoms may appear in an animal when 2 percent of its body weight in either the leafy tops or fleshy root of the plant is eaten. Trembling, convulsions, and falling with the legs becoming rigid are the first symptoms. Difficult breathing and acute pain will generally pass with only mild poisoning, and the animal will return to an upright position within 20 minutes.

What to Do

Supportive treatment is indicated as needed.

Pokeweed

Phytolacca americana*—Phytolaccaceae,* Pokeweed Family

Pokeweed, *Phytolacca americana*

Pokeweed berries

Pokeweed
Up to 9 feet tall

Also Known As

Pigeon Berry, Scoke, Garget

Description

Pokeweed is a distinctive perennial weed, multi-branched, growing 6–9 feet tall. It has a woody stalk and alternating leaves are 5–10 inches long. Racemes of small white or greenish flowers grow at the ends of its red-purple stems, and the fruits, maturing in July–September, are drooping clusters of shiny purple berries.

The ripe fruits are minimally toxic, while the fleshy taproots are the most toxic part of the plant; all parts contain toxic saponins and the alkaloids phytolaccatoxin and phytolaccin, which give a strong odor to the plant. Horses and other animals may eat the plant although it is not very palatable.

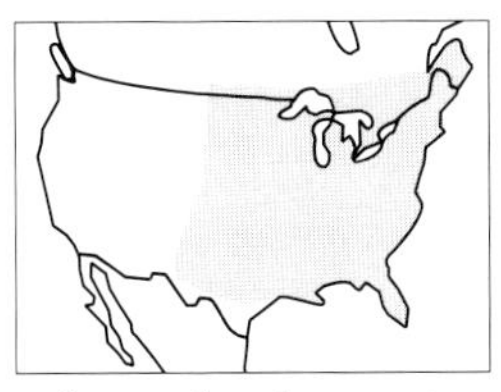

Geographic Distribution

Rich soils and lowlands, clearings, pastures, and neglected areas; in southeastern Canada, eastern United States to Minnesota, and much of Texas. Horsemen in Texas believe pokeweed to be a serious problem in many parts of the state.

Signs of Poisoning

Salivation, colic, muscular weakness and diarrhea (often bloody) develop several hours after moderate quantities of the plant have been eaten. Respiratory failure, anemia, and ulcerative gastritis are symptoms of more severe pokeweed poisoning.

What to Do

Activated charcoal and saline cathartics are indicated to evacuate the intestinal tract. Oils and protectants for the gastrointestinal tract, stimulants, and blood transfusions may be indicated. Prognosis is favorable if treatment is received immediately. Wear gloves when disposing pokeweed from your field, or the toxins may be absorbed through your skin.

Buckwheat

Fagopyrum esculentum—*Polygonaceae,* Buckwheat Family

Buckwheat, *Fagopyrum* species

Buckwheat flowers

Buckwheat
Up to 3 feet tall

Description

Buckwheat has alternate, arrowhead-shaped leaves, jointed stems, and sprays of small, greenish white or pink flowers clustered in racemes. It is toxic to animals when eaten in large quantities.

Fagopyrin is the primary toxin, being present in all parts of the plant. Once absorbed into the blood, the fagopyrin reacts with ultraviolet light in the non-pigmented skin to cause a photosensitization reaction in the skin.

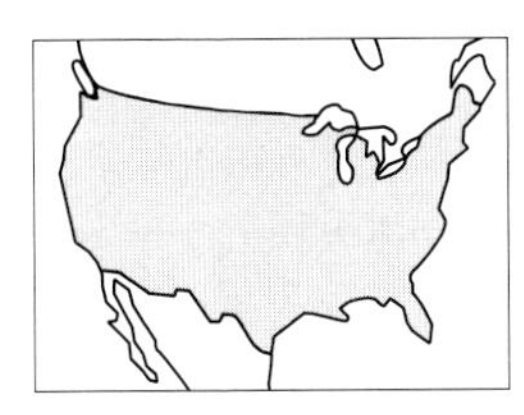

Geographic Distribution

Throughout the United States; cultivated as a cover crop and sometimes as a grain crop for use as flour.

Polygonaceae Family

Signs of Poisoning

Horses with white skin are most severely affected. The skin initially becomes reddened, swollen, and painful. Exposure to sunlight worsens the reaction to the point that the affected skin sloughs, leaving a severely ulcerated lesion. The pigmented areas of the animal are usually unaffected. Animals may try to get into shaded areas to avoid the intense photosensitivity.

What to Do

If animals are promptly removed from the source and kept out of the sunlight, symptoms will subside. Laxatives will assist in removing any residual plant from the gastrointestinal tract. The skin generally heals in one to two months. Permanent scars may remain, however, if an animal's skin was severely affected. Buckwheat should *not* be fed to animals.

Buttercup

***Ranunculus* species**—*Ranunculaceae,* Buttercup Family

Buttercups,
Ranunculus californicus

Buttercup,
Ranunculus bulbosus

Buttercup
Up to 2 feet tall

Also Known As
Crowfoot

Description
Buttercups are perennials or annual herbs ranging from 6 inches to 2 feet tall. They have hollow stems, palmate basal leaves, and smaller, alternate deeply divided stem leaves. The yellow flowers have five petals and five sepals.

Buttercups contain ranunculin, which is converted to the irritant protoanemonin when the plant is chewed. The toxicity varies with the stage of growth, being most toxic when flowering and least toxic when dried. Although animals do not like the bitter taste, some may seek out the plant to eat and show a preference for it, especially if it has been fed to the animal before.

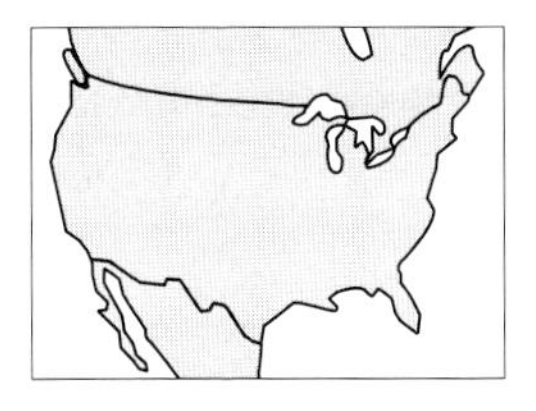

Geographic Distribution
Worldwide, especially in marshy fields and pastures.

Signs of Poisoning
Conditions of poisoning will depend on the age and condition of the animal as well as on growth conditions. Symptoms of poisoning include salivation, depression, blindness, and blood-stained urine. Juice from the plant may cause ulceration of the animal's skin or lips, and salivation, diarrhea, nervousness, and abdominal pain are likely.

What to Do
All buttercups should be removed from the horse's diet. There is no specific treatment, and supportive care is needed.

Larkspur

Delphinium species—*Ranunculaceae,* Buttercup Family

Larkspur, *Delphinium glaucum*

Desert Larkspur, *Delphinium parishii*

Delphinium, Pacific Coast hybrid, *Delphinium* species

Larkspur
Up to 6 feet tall

Also Known As

Poisonweed, Staggerweed

Description

Larkspurs are perennial erect herbs growing from 1 to 6 feet tall depending on the species. Leaves are palmate, often deeply divided, alternate, and larger at the bases of the stem. The characteristic spurred flowers are produced as racemes above the leaves. Although most are blue to purple, white, red, and yellow forms exist.

The toxic principles are polycyclic diterpenoid alkaloids, occurring in highest concentration in the early plant growth and the seeds. Many poisonings occur either early or late in the season. The tall, unpalatable stalk usually deters horses from ingesting many of these plants unless other forage is scarce. Cattle are highly susceptible to poisoning, while horses are more tolerant of the toxins.

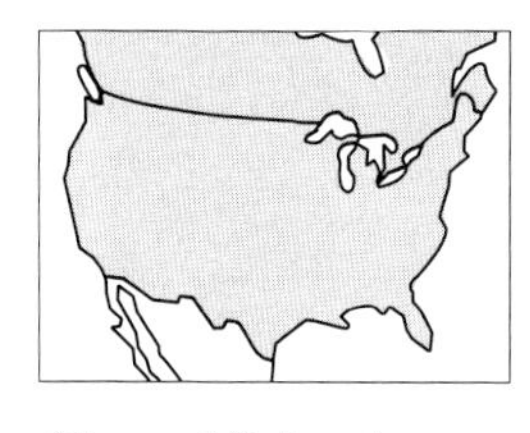

Geographic Distribution

Meadows, woods, and slopes throughout the United States and Canada. Larkspurs tend to cause problems in western North America.

Signs of Poisoning

Stiffness, inability to stand, muscle twitching (especially in the muzzle, shoulder, and flank), abdominal pain, rapid and weak pulse, bloating, and constipation may all be observed either immediately or 24 hours after an animal consumes the plant. Acute poisoning progresses to fatal cardiac and respiratory failure.

What to Do

Avoid stressing the affected animal as this may result in death. Physostigmine is the treatment of choice for larkspur poisoning, and it will reverse some of the life-threatening effects of the alkaloids. It must be given early in the course of poisoning. Activated charcoal with a magnesium sulfate laxative may help to reduce further absorption of the toxins.

Marigold, Marsh

Caltha palustris—*Ranunculaceae,* Buttercup Family

Marsh Marigold,
Caltha palustris
Up to 2 feet tall

Marsh Marigold flowers

Also Known As

Cowslip

Description

The marsh marigold is a succulent plant that grows 1–2 feet high, with a thick hollow stem and very broad, almost heart-shaped leaves. Its shiny yellow 5-petaled flowers bloom April–June, the center of the flower being round with a hairlike surface. Pod-shaped fruits emerge in a whorl arrangement.

Marsh marigolds are apparently only mildly toxic at their full growth stage, while young or dried plants, sometimes found in hay, are even less poisonous and often have no toxicity at all. The plant's toxic factor seems to stem from its anemonin content. Cattle, sheep, and horses have been poisoned by eating fresh tops.

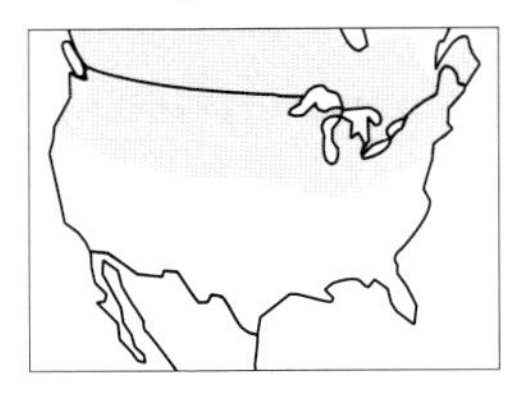

Geographic Distribution

Swamps, marshes, and along waterways; across Canada and the northern United States.

Signs of Poisoning

Juice from the plant may cause ulceration of the animal's skin or lips, and salivation, diarrhea, nervousness, and abdominal pain.

What to Do

No treatment has been noted. Avoid grazing horses on the marigold whenever possible.

Monkshood

Aconitum napellus—*Ranunculaceae,* Buttercup Family

Cultivated Monkshood, *Aconitum napellus*

Monkshood flower, *Aconitum napellus*

Monkshood
Up to 5 feet tall

Also Known As

Crowfoot

Description

Monkshood is a summer-flowering herbaceous perennial, similar in characteristics to delphinium. It is a slender plant growing 3–5 feet tall, with alternate, deeply cut, 2- to 4-inch-wide palmate leaves. Its hood- or helmet-shaped flowers are 11/4–11/2 inches high and appear in racemes at the top of the stalk; these are usually a distinct blue or purple, but may occasionally be white, peach, or pink. They lack the characteristic spur of the delphinium's flowers. The dried seed pods contain numerous tiny seeds.

All parts of the plant are toxic, especially its leaves and tuberous roots. Aconitine and other related alkaloids are the toxic principles. Only .075 percent of a horse's weight in root (less than 1 pound) has been lethal.

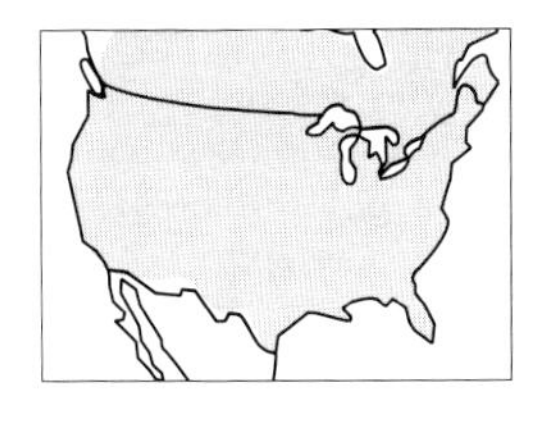

Geographic Distribution

Wild and cultivated, in shaded moist soils, throughout Canada and the United States except for regions of extreme temperatures.

Signs of Poisoning

Poisoned horses may exhibit weakness, restlessness, irregular heartbeat, salivation, and prostration. Livestock commonly experience belching, bloating, and constant swallowing. Death is common within hours of ingestion.

What to Do

There is no specific treatment. Affected animals should be given activated charcoal via stomach tube if the plant has been recently eaten. Intravenous fluid therapy may be necessary to help manage the cardiovascular effects of the toxins.

Foxglove

Digitalis purpurea—*Scrophulariaceae,* Snapdragon Family

Foxglove, *Digitalis purpurea*

Foxglove flowers

Foxglove
Up to 5 feet tall

Description

Foxglove is noted for its showy, slender, purple, pink, red, or white tubular flowers with spots that resemble the part of a glove where the fingers enter. Leaves are alternate, hairy, and slightly toothed.

Horses and cattle rarely eat the plant. Its toxic principles are saponins, alkaloids, and the cardiac glycosides digitoxin, digitalin, and digoxin. When consumed, it takes only a few hundredths of a percent of an animal's weight to be fatal. Toxicity is not affected by drying or aging.

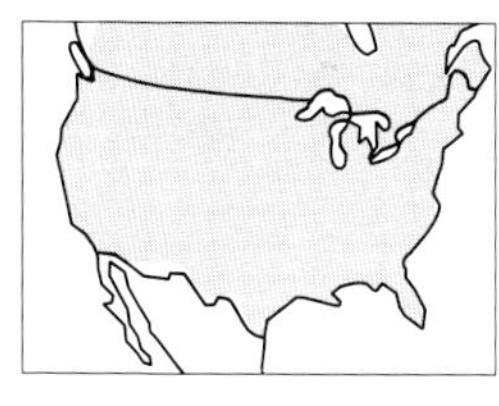

Geographic Distribution

Cultivated in flower gardens throughout the United States; wild in western states and Canada, commonly on farmland.

Signs of Poisoning

Colic, bloody feces, poor appetite, pain, frequent urination, irregular heartbeat and pulse, and possible convulsions are symptoms in the horse prior to death.

What to Do

No treatment has been noted other than symptomatic. Activated charcoal and saline laxatives are appropriate following recent ingestion of the plants. Intravenous fluid therapy, monitoring of serum potassium levels, and other supportive therapy will aid in recovery.

Indian Paintbrush

Castilleja species—*Scrophulariaceae,* Snapdragon Family

Indian Paintbrush,
Castilleja rhexifolia

Indian Paintbrush,
Castilleja cottinea

Indian Paintbrush
Up to 2 feet tall

Description

Indian paintbrush is a 1- to 2-foot-high annual with a slender taproot. Its leaves are alternate, and the showy flowers are white or greenish yellow and hidden within red, yellow, or green leaves (bracts).

The plant is known to accumulate selenium, and grazing animals ingesting the plant can be poisoned.

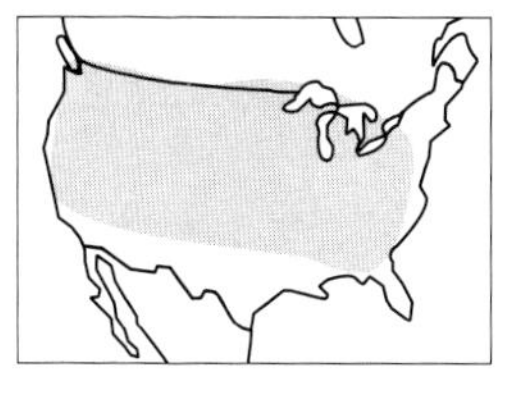

Geographic Distribution

Meadows and rangelands; damp sandy soils; across the United States. as far south as northern Florida and Oklahoma.

Signs of Poisoning

Animals that eat the plant may develop chronic selenium poisoning. Symptoms range from loss of long mane and tail hair, dullness, rough coat, cracking of hooves, stiff joints, and lameness to staggering, aimless wandering, and eventually anorexia, emaciation, and depression. Coma and death may occur in acute poisoning. (*See* Prince's Plume)

What to Do

Remove the horse promptly from the hay or field. Once signs of chronic selenium poisoning are present, only symptomatic treatment is possible. Meticulous hoof care will help prevent loss of the hoof wall.

In selenium rich soils of North America, it is important to feed a balanced ration with adequate sulfur and copper levels to counteract the high selenium in the plants. Alfalfa with high levels of sulfur-containing amino acids is an ideal food source. Dried molasses and bran are good sources of copper, sulfur, and trace minerals, but the extra phosphorus in bran must be carefully balanced with a calculated amount of calcium.

Buffalo Bur

Solanum rostratum—Solanaceae, Nightshade Family

Buffalo Bur,
Solanum rostratum

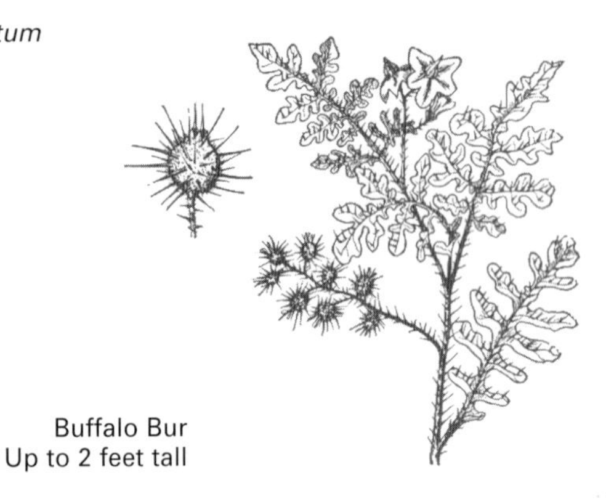

Buffalo Bur
Up to 2 feet tall

Weeds & Wildflowers

Also Known As

Kansas or Texas Thistle

Description

The buffalo bur is an annual plant that grows about 2 feet tall and is covered in prickles, with wide, deeply lobed spiny leaves of irregular lengths and yellow flowers. Its distinct fruit is an oval spiny bur. As a member of the nightshade family, it contains steroidal solanine alkaloids and glycoalkaloids. It is toxic to all animals.

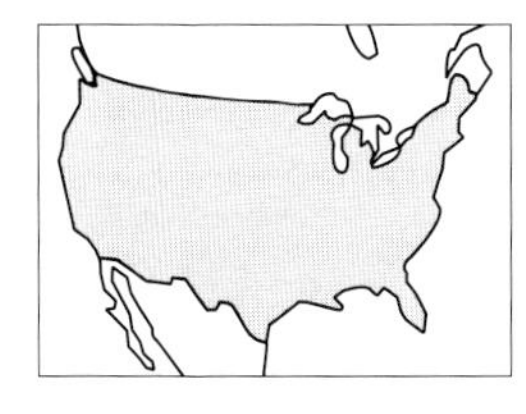

Geographic Distribution

Neglected and overgrazed areas of the United States.

Signs of Poisoning

Buffalo bur causes mechanical injury to the mouth and esophagus and gastrointestinal tract. Neurological disorders including tiredness, trembling, and bloating, and congestion of the lungs, heart, liver, and spleen may occur if the young green plants are eaten in quantity.

What to Do

There is no specific treatment, and successful supportive care must be initiated early in the course of poisoning. Activated charcoal and laxatives may help decrease toxin absorption. Intravenous fluid therapy and means to support respiration should be available. Physostigmine may be helpful in severe cases.

Jimsonweed

***Datura* species**—*Solanaceae,* Nightshade Family

Jimsonweed,
Datura stramonium
Up to 5 feet tall

Datura stramonium

Jimsonweed with trumpet flowers and seed pod

Also Known As

Datura, Jamestownweed, Apple of Peru, Thornapple

Description

There are more than a dozen different species of *Datura*, known by a variety of names throughout the world. Most of them grow up to 5 feet in height and have large wavy leaves, fragrant trumpet-shaped flowers 3–6 inches long in white to violet, and 2-inch spiny fruit pods. Horses and livestock will eat this plant only when other food is not available, as the green part of the plant has a strong odor.

All these plants contain a number of solanaceous alkaloids such as solanine, hyoscyamine, and hyoscine, which interfere with digestion and the nervous system when ingested. All parts of the plant are highly toxic. Although used for medicinal purposes in the past, they are lethal when consumed in more than minute amounts.

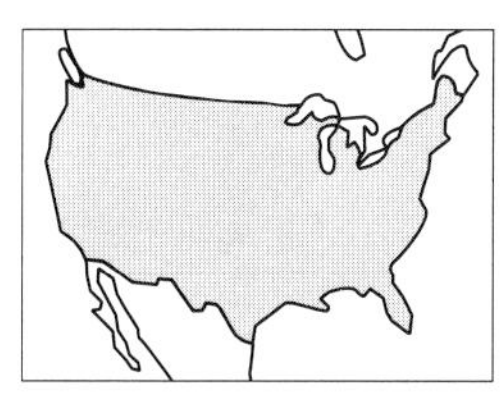

Geographic Distribution

Fields and neglected places, especially on rich bottom soils, throughout the United States. Some species are cultivated as an ornamental and may be purchased in garden shops.

Signs of Poisoning

Symptoms occur within minutes or a few hours after ingestion. They include excitement and then sudden depression, colic, diarrhea, thirst, convulsions, subnormal temperatures, dilated pupils, coma, respiratory paralysis, and quick death.

What to Do

There is no specific treatment, and successful supportive care must be initiated early in the course of poisoning. Activated charcoal and laxatives may help decrease toxin absorption. Intravenous fluid therapy and means to support respiration should be available. Physostigmine may be helpful in severe cases.

Nettle, Bull

Solanum carolinense—*Solanaceae,* Nightshade Family

Horse Nettle, *Solanum carolinense*

Weeds & Wildflowers

Bull Nettle
Up to 2 feet tall

Also Known As

Horse Nettle

Description

This erect perennial reaches 6 inches to 2 feet in height, with alternate, widely toothed leaves. Leaf veins are spiny on the undersides, and the stems also bear spines. White or violet flower clusters and yellow berries 1/2 inch in diameter may also be seen.

The toxic principle is solanine; toxicity increases dramatically in fall when the plant reaches maturity. Even in winter, the dried berries are toxic.

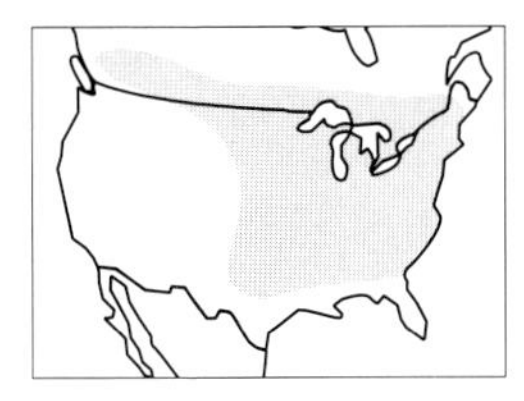

Geographic Distribution

Neglected areas, from southern Canada south through eastern and central United States to Texas.

Signs of Poisoning

Anorexia, emaciation, rough coat, constipation, intestinal lesions, and inflammation of the mouth have been recorded in livestock.

What to Do

There is no specific treatment, and successful supportive care must be initiated early in the course of poisoning. Activated charcoal and laxatives may help decrease toxin absorption. Intravenous fluid therapy and means to support respiration should be available. Physostigmine may be helpful in severe cases.

Nightshade, Common

Solanum americanum—*Solanaceae,* Nightshade Family

Common Nightshade, *Solanum americanum*

Nightshade flowers

Common Nightshade
Up to 3 feet tall

Also Known As

Deadly or Black Nightshade, *Herba mora negra*; American Nightshade

Description

Common nightshade is an annual plant that grows to a maximum of 3 feet tall. It has translucent, pointed oval leaves, about 4 inches long. Its white to purplish drooping star-shaped flowers, with five petals and five stamens, typical of the plants in the nightshade family, bloom year round. The fruits are small, shiny black berries.

The toxic principle is glycoalkaloid solanine. The toxicity levels in the plant vary with the climate and stages of growth, with the highest concentration probably being in the unripe berries. Animals may be attracted to the green plant for grazing. All nightshades should be considered poisonous.

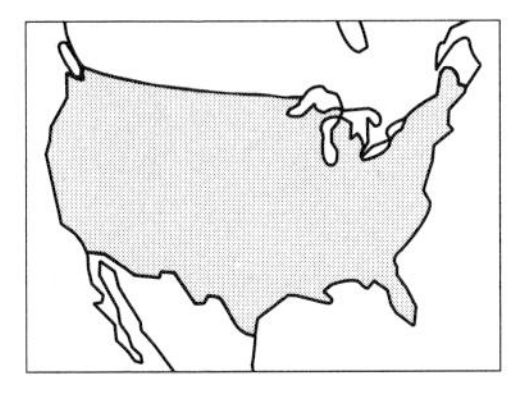

Geographic Distribution

Along fencerows, in neglected areas and disturbed soils; throughout the United States and especially among grain crops, where it may become mixed in with the harvest.

Signs of Poisoning

Animals that ingest the nightshade plants will show signs of neurological and gastrointestinal disorder, tiredness, muscle twitching, bloating, and congestion in the lungs, heart, and spleen.

What to Do

There is no specific treatment, and successful supportive care must be initiated early in the course of poisoning. Activated charcoal and laxatives may help decrease toxin absorption. Intravenous fluid therapy and means to support respiration should be available. Physostigmine may be helpful in severe cases.

Nightshade, Silverleaf

Solanum elaeagnifolium*—Solanaceae,* Nightshade Family

Silverleaf Nightshade, *Solanum elaeagnifolium* Up to 3 feet tall

Silverleaf Nightshade flowers and flower buds

Also Known As

Trompillo, White Horse Nettle

Description

Silverleaf nightshade stands 1 to 3 feet tall, has white hairy leaves and stems, and star-shaped blue-violet flowers. A hard yellow-orange berry grows in the center of the flower (*see* Common Nightshade).

This species has caused heavy losses in livestock. Solanine is in the fruit, which is especially toxic when ripe, unlike other nightshades, whose unripe berries are most toxic. Only a few ounces of the fruit can cause severe poisoning and death.

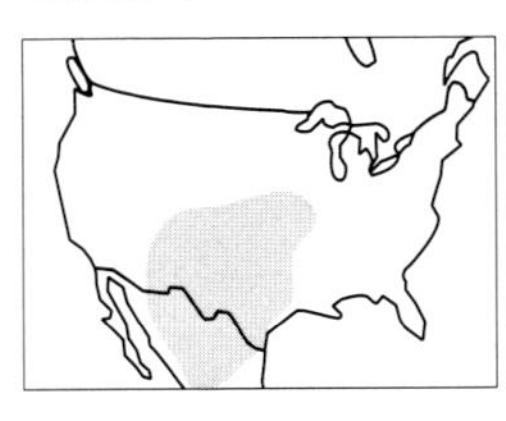

Geographic Distribution

Disturbed soil, open woods, and prairies; southwestern United States from Mexico north to Missouri.

Signs of Poisoning

Animals that ingest the plant will show signs of neurological and gastrointestinal disorder, tiredness, muscle twitching, bloating, and congestion in the lungs, heart, and spleen. Respiratory paralysis precedes death.

What to Do

There is no specific treatment, and successful supportive care must be initiated early in the course of poisoning. Activated charcoal and laxatives may help decrease toxin absorption. Intravenous fluid therapy and means to support respiration should be available. Physostigmine may be helpful in severe cases.

Tobacco

Nicotiana species—*Solanaceae,* Nightshade Family

Tobacco, *Nicotiana* species

Tobacco flowers

Tobacco, *Nicotiana* species

Tobacco can grow as a shrub up to 4 feet tall or as a tree up to 18 feet tall.

Tobacco

Description

There are three naturally ocurring species. Wild tobacco, *N. attenuata*, is an herbaceous, branching plant growing 1–4 feet tall with hairy, sticky stems. It has alternate leaves, 11/2–4 inches long, and white, tubular, five-petaled flowers. Tree tobacco, *N. glauca*, is a tall, slender, evergreen shrub or tree, 6- to 18-feet tall. Its flowers are the same as above, but yellow. Desert tobacco, *N. trigonophylla*, is a slender, herbaceous annual growing 1–3 feet tall. Its stems are hairy and sticky, and its alternate leaves are attached with clasping wings. It flowers as above in either white or yellow.

Not only is the nicotine in the plant toxic, but a variety of other alkaloids contained in the leaves and flowers are suspect. These plants cause severe poisoning in horses and other animals when ingested. Animals with free access to the plants in the field will eat them voluntarily when other food is scarce.

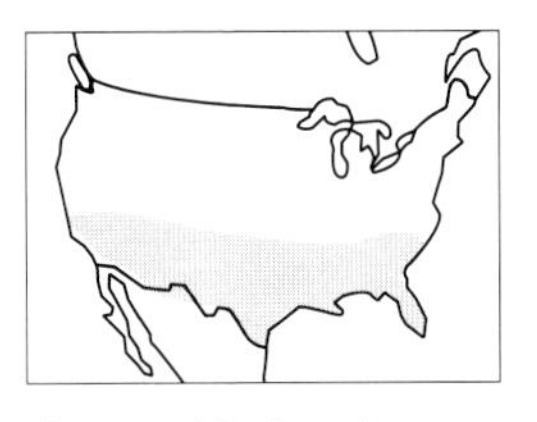

Geographic Distribution

Dry, sandy soils and neglected areas, in western and southwestern United States. *Nicotiana* tobacco is cultivated in eastern and southern states for commercial use. It, too, is toxic.

Signs of Poisoning

Both processed tobacco and plants in the field cause a variety of symptoms. Symptoms of poisoning include severe muscle twitching, rapid weakened pulse, abdominal pain, respiratory difficulty, and diarrhea. These symptoms will appear shortly after ingestion, and death may occur minutes or days later, depending on the animal and the amount of the plant consumed.

What to Do

There is no specific treatment, and poisoned animals should be given supportive therapy, including activated charcoal orally, intravenous fluids, and artificial respiration when needed.

African Rue

Peganum harmala—*Zygophyllaceae,* Caltrop Family

Rue, *Ruta graveolens* Up to 10 inches tall

Rue, *Ruta* species

Weeds & Wildflowers

Similar Species

Mexican Rue, *Peganum mexicanum*

Description

African rue can be recognized by its many branches, growing up to 10 inches tall, with alternating segmented leaves and white or yellowish 5-petaled flowers. Small fruit capsules are also present. The Mexican rue is similar.

Both species contain the alkaloids vasicine, harmaline, harmine, and harmalol, and the seeds are most toxic. The most dangerous periods are in spring, when new shoots are coming up, or under drought conditions, when there is little else to eat. The full grown plant is unpalatable but will be eaten if better forage is unavailable. Horses are potentially susceptible.

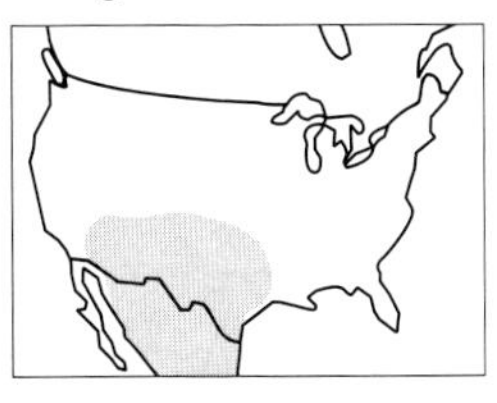

Geographic Distribution

On poor and overgrazed soils, in arid and semi-arid places; southwestern United States and Mexico.

Signs of Poisoning

Ingestion results in rigidity of the body, lack of coordination, shaking, frequent urination, listlessness, severe gastrointestinal distress, and hemorrhages in the liver. Excessive salivating is reported as a common finding in poisoned animals.

What to Do

There is no treatment for African rue or Mexican rue poisoning. Horses should be given a good supply of feed and water and be removed from exposure to these plants.

Arrowgrass

Triglochin maritima—*Juncaginaceae* Family

Arrowgrass, *Triglochin maritima* Up to 3 feet tall

Arrowgrass flower spikes

Grasses & Horsetails

Similar Species

Triglochin palustris

Description

Arrowgrass can be recognized by its short basal stem and clumps of grasslike leaves that are round with a flattened side. Flowers are small and greenish, produced on a tall spike. The fruits, composed of up to 6 capsules, enable recognition of the plant in a hayfield.

Arrowgrass has prussic acid in its leaves and can have a high cyanide content, depending upon the location and conditions under which it is grown. It is recorded to be potentially lethal to animals at .5 percent of their body weight.

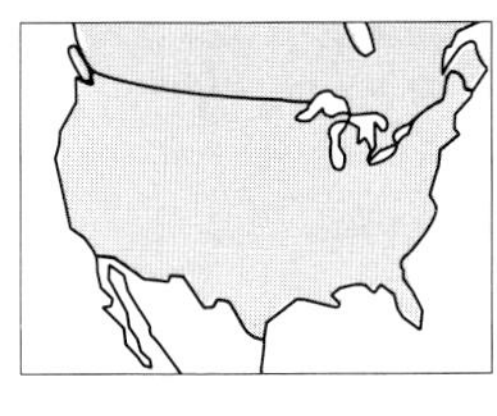

Geographic Distribution

Damp, alkaline soils, shorelines, bogs, and salt marshes; throughout the United States and Canada.

Signs of Poisoning

Symptoms of arrowgrass poisoning are typical of cyanide poisoning: excitement, rapid respiration, weakened pulse, tachycardia, salivation, voiding of urine and feces, staggering, collapse, bright red mucous membranes, convulsions, and death.

What to Do

Administer IV solution of sodium nitrite and sodium thiosulfate. Although the plant seems to lose some of its toxicity when dried in hay, it should still be avoided.

Dallis Grass

Paspalum dilatatum—*Poaceae,* Grass Family

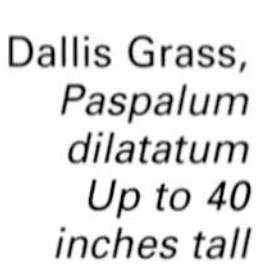

Dallis Grass, *Paspalum dilatatum* *Up to 40 inches tall*

Dallis Grass stem with seeds

Similar Species

Rye Grass, *Lolium perenne* (with *Ergot* fungus, *Claviceps* species)

Description

Dallis grass is a common perennial grass that may grow as tall as 40 inches under the right conditions. The blades are flat, coarse, and have pointed ends. Little hairs grow at their base and dozens of tiny oval seeds grow up the stem. Rye grass is a coarse green annual or perennial grass with a spiked end, similar to dallis grass, but it grows only to about 25 inches tall. Its seeds, which grow up the sides of the plant, are somewhat flatter and less sparse than those of the dallis grass. Frequently, a parasitic fungus invades the flower heads, producing "honey dew." Insects are attracted to the secretion and help in transmitting the fungus. This fungus produces lysergic acid derivatives, ergotamine, and ergotoxine, affecting animals that ingest it. The endophytic fungus (*Acremonium lolii*) that invades rye grass produces a tremorgenic toxin that induces muscle tremors.

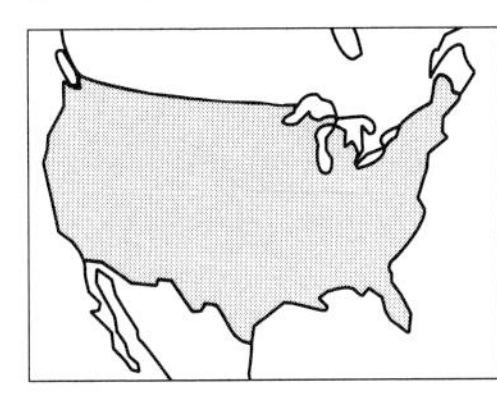

Geographic Distribution

Open fields with dry, moist, or sandy soils; throughout the United States. Rye grass is a frequent choice for planting in yards during winter.

Signs of Poisoning

Symptoms include nervousness; trembling, staggering; abortion; convulsions; blood vessel restriction causing nerve damage in the tail, ears, and other limbs; lameness; and gangrene. These symptoms occur within several days to several weeks of ingestion. Cattle are more commonly involved in dallis or rye grass poisoning, but other animals, including horses, may also be susceptible.

What to Do

There is no treatment except to change the animal's diet. Always keep dallis and rye grasses mowed in your pasture, and never feed horses grass clippings. It is also important to avoid overgrazing rye grass, as the fungus exists near the base of the plant and will be eaten if animals are grazing close to the ground.

Fescue

Festuca species—*Poaceae,* Grass Family

Tall Fescue, *Festuca* species Up to 4 feet tall

Tall Fescue flower spike

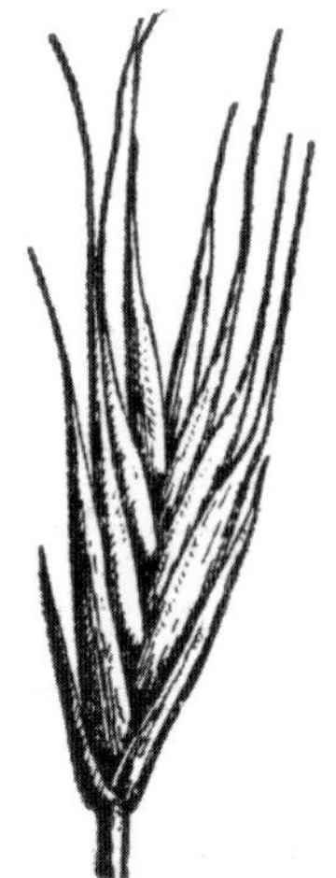

Tall Fescue

Description

Chewings fescue, *Festuca rubra,* grows to 3 feet tall and has round wiry leaves that grow in tufts at the base of the plant. Nematode galls are often found in the seed and have produced toxic symptoms in horses after they ingest either the seed or the grass. Coryne toxins are the toxic principle.

Tall fescue, *Festuca arundinacea,* is a drought-resistant, coarse perennial grass that thrives in wet areas and is often grown for forage. It has a long, flat, ribbed, dark green blade, and can grow up to 4 feet. It bears many small flowers on 1-foot spikes. The plant contains alkaloids, perloline, and halostachine. An endophyte fungus is known to infect fescue and is important to the development of toxicity to animals grazing the grass.

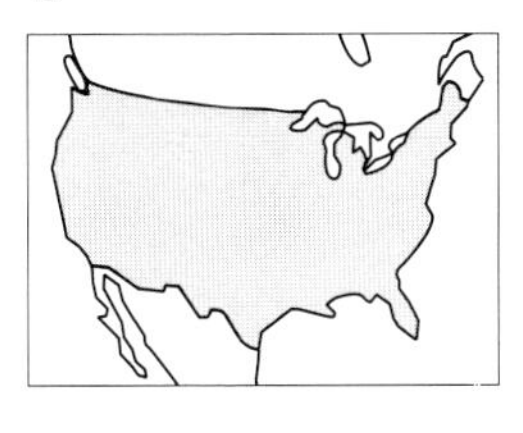

Geographic Distribution

Chewings fescue grows in dry or rocky soils, on lawns throughout the United States. Tall fescue grows in wet areas, throughout the United States.

Signs of Poisoning

With chewings fescue poisoning, muscular trembling, ataxia, staggering and falling, abortions, and death have all been noted, with degeneration of liver and kidneys in chronic cases.

Tall fescue poisoning occurs after several days to several months of grazing endophyte-infected fescue. Symptoms may vary depending upon the time of the year: In winter, lameness, diarrhea, anorexia, rough hair coat, and possible gangrene of the tail, hooves, and ears may appear. Poor growth rates and weight loss may occur. In summer, animals may have elevated temperatures, and females have little milk for their young. Stillbirths, abortions, prolonged gestations, retained placentas, and infertility are frequent in mares. Foals may have very long hooves if they survive birth due to the prolonged gestation.

What to Do

No treatment has been noted. Do not allow animals to graze where these grasses are grown. Endophyte-free fescue should be used to reseed pastures.

Johnson Grass

Sorghum halepense—*Poaceae,* Grass Family

Johnson Grass, *Sorghum halepense*

Sudan Grass, *Sorghum vulgare*

Sorghum
Up to 8 feet tall

Similar Species

Columbus Grass, Sorghum, *Sorghum* species; Sudan grass, *Sorghum vulgare*

Description

Johnson grass is a coarse perennial grass with large runners (rhizomes) and topped with clusters of flowers. Sudan grass is an annual, erect plant 6–8 feet high with a terminal florescence resembling corn. Sudan grass and its hybrids are often grown as a forage crop for horses and cattle. Animals consuming them in either fresh or dried form may suffer cyanide poisoning.

Both Johnson and Sudan grasses may contain hydrocyanic acid (prussic acid) and sometimes toxic levels of nitrates. Toxicity is highest in young plants and lowest when the plant is yellow, more than 2 feet tall, and forming fruiting heads. The levels of cyanide increase when the plant is stressed, for example during drought or frost. Leaves have a higher concentration of cyanide than stems.

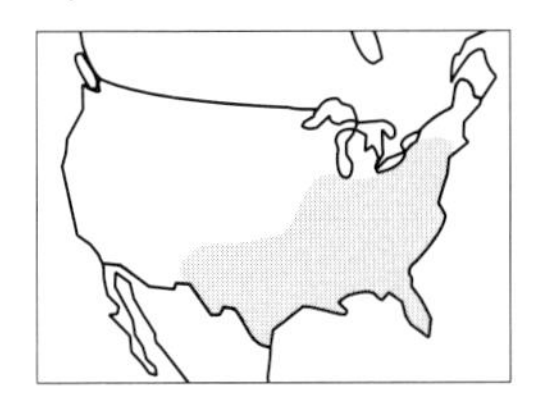

Geographic Distribution

Open fields and neglected areas, throughout the southern United States and north to Iowa and New York.

Signs of Poisoning

Horses may suffer from acute and chronic cyanide poisoning when eating sorghums. Symptoms of acute poisoning include excitement, rapid respiration, weakened pulse, tachycardia, salivation, voiding of urine and feces, staggering, collapse, bright red mucous membranes, convulsions, and death. Horses consuming sorghum hay for long periods may develop chronic cyanide poisoning that causes nerve degeneration in the hind legs, urinary tract, bladder, and rectum. Affected horses show weakness and an unsteady gait of the hind legs. They also develop urinary incontinence and an atonic rectum that becomes impacted with feces. Recovery from chronic cyanide poisoning is unlikely as nerve degeneration is permanent. Pregnant mares may abort or give birth to deformed foals.

What to Do

Treat for cyanide poisoning (IV solution of sodium nitrite and sodium thiosulfate) or nitrate/nitrite poisoning (IV solution with 1 percent methylene blue). Avoid frost-damaged plants for animal forage. If feeding sorghum hay to horses be sure it is made from cyanide-free varieties of sorghum.

Kleingrass

Panicum coloratum—*Poaceae,* Grass Family

Kleingrass,
Panicum vitgarum
Seed heads grow up to 4 feet tall.

Kleingrass stems and blades

Similar Species

Panicum vitgarum

Description

Kleingrass is a perennial with narrow blades and looks similar to coastal grass (the hay of choice in Texas). Look for bunches of small seeds at the tops to help identify it. Kleingrass grows up to 4 feet tall and bears small spiklets on its tops at maturity.

The toxic principle is believed to be saponin, but it is not found in the same quantities in all plants. Although the grass has a strange smell, it may be eaten by hungry horses when no other forage is available. Kleingrass hay is also toxic.

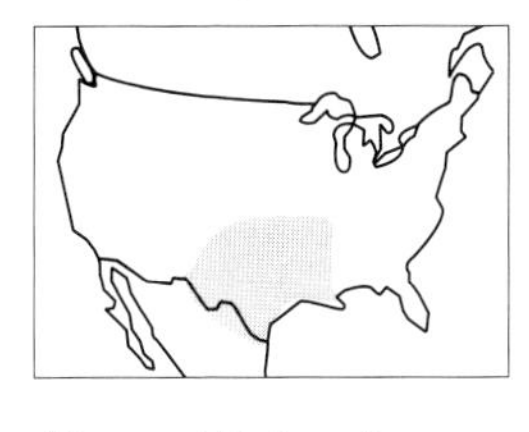

Geographic Distribution

Texas. Kleingrass originally came from South Africa and was introduced to Texas through Texas A and M University.

Signs of Poisoning

Cattle do not seem to be affected by this grass, but other livestock and horses are very susceptible. Horses that eat kleingrass will develop liver disease and photosensitization around the coronary band.

What to Do

No treatment has been noted. Horses should be removed from the kleingrass source, put on good quality hay, and kept out of the sunlight.

Squirreltail Grass

Hordeum jubatum—*Poaceae,* Grass Family

Squirreltail Grass,
Hordeum jubatum
Up to 18 inches tall

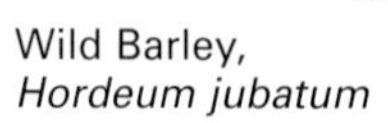

Wild Barley,
Hordeum jubatum

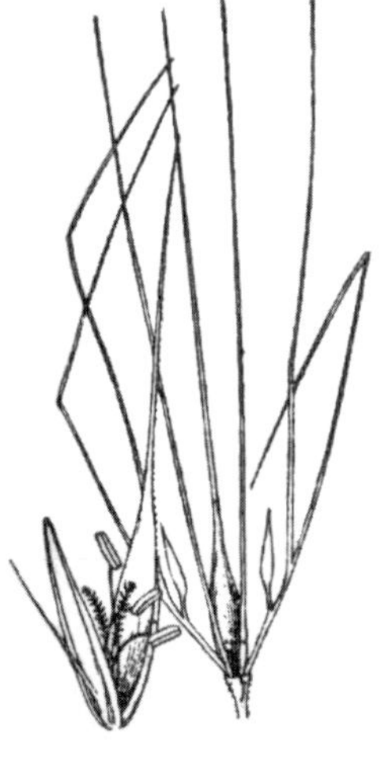

Squirreltail Grass

Also Known As

Foxtail Grass, Wild Barley

Description

Squirreltail grass grows tall with wiry bristles and a flowering spike with tiny teeth; these teeth can penetrate flesh and hook onto it so it cannot be removed. Horses and livestock may be injured from this plant when grazing or eating poor quality hay.

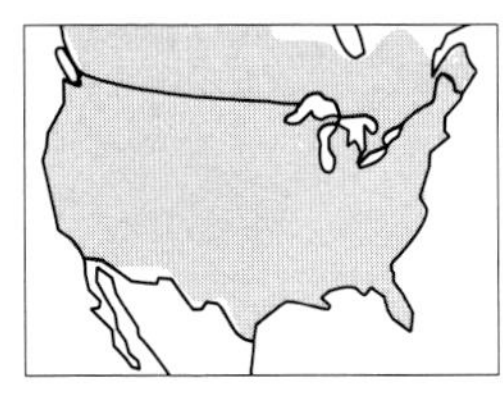

Geographic Distribution

Throughout the United States and north into Canada.

Signs of Poisoning

The grass may pierce the skin on the animal's ears, neck, face, or mouth, causing abscesses, ulcers, possible blindness, and the inability to eat. It may cause colic and impaction in horses.

What to Do

Colic treatment or surgery and other treatment may be necessary. Consult your veterinarian.

Avoid overgrazing, which allows this undesirable grass to become established.

Yellow Bristle Grass

Setaria Lutescens*—Poaceae,* Grass Family

Foxtail Grass, *Setaria viridis*

Foxtail Grass flower

Yellow bristle grass seed heads grow to 40 inches tall.

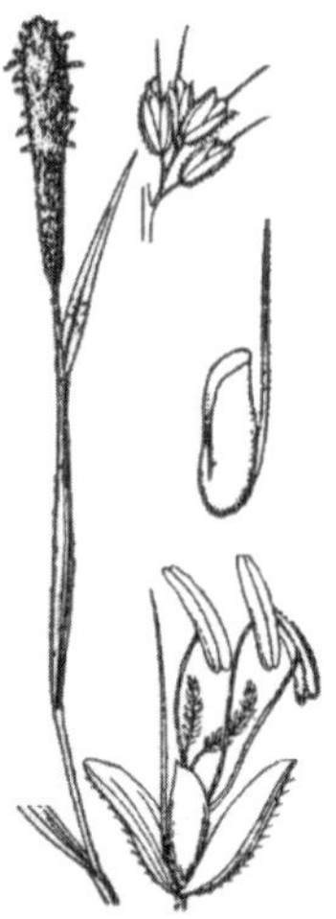

Foxtail Grass

Also Known As
Foxtail Grass, Pigeon Grass

Description
Yellow bristle grass does not contain toxins, but it is a poor forage for animal consumption. It has little spikes and wiry bristles with tiny barbs on the ends that cause mechanical injury to an animal's oral tissues.

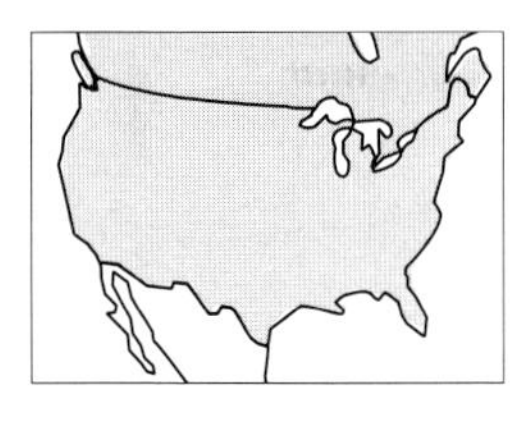

Geographic Distribution
Roadsides and range areas, throughout the United States and Canada.

Signs of Poisoning
While being both chewed and digested, the barbed bristles cause ulcers in the mouth and digestive tract. Horses are especially susceptible to mechanical injury because they have softer oral mucous membranes than livestock.

What to Do
Oral ulcers should be explored for the embedded grass awns. The awns must be removed before healing will take place.

Horsetails

Equisetum arvense—*Equisetaceae,* Equisetum Family

Horsetails,
Equisetum arvense

Horsetails—fertile stalks

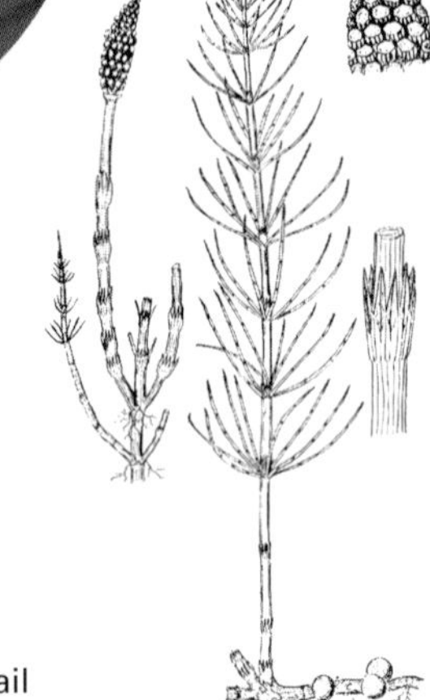

Horsetail
Up to 3 feet tall

Also Known As

Foxtails, Scouring Rushes

Description

Horsetails have windswept-looking whorls of thin, grasslike leaves at the ends of the many tall, green, segmented hollow stems, resembling a horse's tail. They are reported to be poisonous to all classes of livestock, although horses appear to be most susceptible.

The plant contains toxic aconitic acid, palustrine, and thiaminase. Horses have shown various degrees of poisoning after consumption, and young horses are more likely to succumb than older horses. Toxicity is higher in green plants than in aged plants.

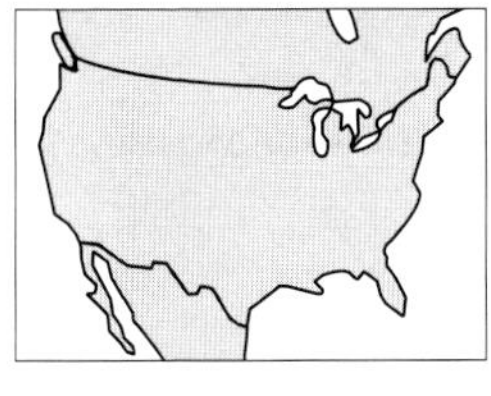

Geographic Distribution

Worldwide; in moist fields, roadsides, and drainage areas, frequently in sandy soil but also in gravel and along waterways.

Signs of Poisoning

Hay containing the horsetail plant fed for a period of 2 weeks has produced symptoms of ill thrift, weakness, and staggering. Sometimes trembling, muscular rigidity, diarrhea, rapid pulse, and cold extremities are also noted. Appetite generally stays the same, and coma precedes death if the animal's food is not changed.

What to Do

Specialized blood work can indicate thiamine deficiency, and massive doses of thiamine given early in the course of poisoning are beneficial. All sources of horsetails should be removed from the horse's diet.

Toxic Suspects

Besides the plants listed in the preceding sections of this book, several more are potentially dangerous to horses. For some of these, horse poisonings have been rare. For others, only livestock poisonings—and not specifically horse poisonings—have been documented. As it may be better to err on the side of caution, they are listed here for your information.

American Bittersweet

Celastrus scandens—Celastraceae, Staff Tree Family—Common in a variety of soils in woods and around fencerows and thickets, usually from Texas east to Georgia, and north into Canada.

Bittersweet is a woody ornamental vine that climbs to 20-foot heights. It has alternate, simple leaves and numerous 5-petaled greenish yellow flowers on the ends of its branches in May and June. The 1/3-inch-thick stems look wrinkled and leathery and may be green, gray, or brown with a white core. Orange to yellow fruits hang in clusters, ripening in August–October. Horses usually avoid this plant if other forage is available. Neither toxic principle nor treatment is known.

Juniper

Juniperus species—Cupressaceae, Cypress Family—Scattered throughout North America, in poor soil areas where other trees may not do well, such as on dry, rocky western slopes.

Junipers are well-known evergreens of many different shapes and sizes, from ground covers to 50-foot-tall trees. They shed a pollen known to cause hay fever, and their cones—resembling blue or brown berries—are toxic. The short, sharp needles covering all the branches and an astringent resin in the fruit make this tree unpalatable for livestock and horses, so it will be consumed only if an animal is extremely hungry. The berries cause severe stomach and intestinal upset, and the needles can cause mechanical injury. No treatment has been noted.

Coyotillo

Karwinskia humboldtiana—Rhamnaceae, Buckthorn Family—Found in dry soils and hillsides of Texas, southern California, Mexico, and Central America.

The coyotillo is a small tree or shrub that grows up to 24 feet. Bark at its base is gray and smooth, while the branches are reddish brown. Oblong leaves with raised veins and black spots grow opposite in pairs to about 2 inches long. Small berries

(1/4–1/2 inch) containing an oily seed appear in October and may be red, brown, or black. Berries, seeds, and leaves are poisonous to all animals, but horses are rarely affected.

When ingested, the oil in the seeds causes paralysis in the extremities; this nerve damage may result in death. Symptoms usually do not appear immediately as the toxin must accumulate to reach its toxic threshold. Other signs include depression, poor condition, trembling, and staggering. Treatment may be given too late if symptoms are not immediately recognized. Only mildly affected animals will recover with intensive nursing as no treatment has been noted.

Day-Blooming Jessamine

Cestrum diurnum—Solanaceae, Nightshade Family—Along roadways, woods, and neglected areas; southern United States from Florida to Texas; ornamentals in warm climates.

There are three different species of jessamine—day-blooming, night-blooming, and willow-leaved jessamine—that are toxic to horses. All have simple alternate leaves about 1-inch long, with fragrant, showy, trumpet-shaped flowers growing in clusters. *C. diurnum* has white flowers and black berries. *C. nocturnum* has greenish yellow flowers and white berries, particularly fragrant at night. *C. parqui* has yellowish green flowers and purplish brown berries. The small berries are much more toxic than the leaves. Solanine and other tropane-related alkaloids are the toxic principles.

Symptoms after ingestion of jessamines include gastroenteritis, blood in the feces, nervousness, muscular twitching, salivation, difficulty breathing, and paralysis. Death may occur shortly after ingestion. As with other solanine-containing plants, there is no specific treatment. Activated charcoal, laxatives, and supportive intravenous fluids are indicated. Animals surviving for 24 hours usually recover.

Texas Nightshade

Solanum triquetrum—Solanaceae, Nightshade Family—On fences and trees, in west Texas and Mexico.

Texas Nightshade is a green viny perennial. It has few small 5-petaled flowers, which may be white, purple, or yellow. The fruits are red berries about 1/3-inch wide with flat seeds. Leaves are simple and triangular, alternating up the thin stem. Stems grow to 5 feet long and tend to spread onto nearby plants.

The plant contains solanine glycoalkaloids, and all parts are reported to be poisonous. Ingestion causes weakness, trembling, nausea, either constipation or diarrhea, and finally death. As with other nightshades, there is no specific treatment, and successful supportive care must be initiated early in the course of poisoning. Activated charcoal and laxatives may help prevent further toxin absorption. Intravenous fluid therapy and means to support respiration should be available. Physostigmine may be helpful in severe cases.

Whitebrush

Aloysia lycioides—Verbenaceae, Verbena Family—On rocky hillsides in desert areas of the southern United States.

Whitebrush is a flowering shrub that grows 3 to 10 feet tall. Its leaves are 1/4- to 1-inch long and pale on the underside. The flowers are small, white or purple, and very fragrant. The toxic principle is yet unknown, but horses and donkeys left to forage on their own for food in Arizona, Texas, and Mexico have been poisoned in the past. Horses have shown poor endurance, lameness, sweating, nervousness, weakness, and finally death. These symptoms occur over the span of 11/2 months as the animal grazes on whitebrush. Brain and kidney damage can also occur. Death occurs after a week of nervous symptoms.

Poisoning from whitebrush is now a rarity because horses will not eat it unless they are extremely malnourished or there is little other forage. There is no known treatment.

Inkweed

Drymaria pachyphylla—Caryophyllaceae, Pink Family—In alkaline soils, overgrazed areas of Texas and the Southwest.

This grayish looking summertime plant has numerous branches that radiate on the ground from a central root. It has small, single white flowers with 5 petals, and its immature fruit sometimes gives off a purplish juice. It is toxic to all livestock but eaten only in droughts, when other forage is unavailable.

Symptoms of poisoning appear within 2–24 hours of ingestion. Loss of appetite, diarrhea, restlessness, depression, and poor appearance may soon be followed by coma and death. Recovery is possible only if the animal is removed from the source before ingesting more than .5 percent of its body weight.

Bagpod

Sesbania vesicaria—Fabaceae, Pea Family—In neglected and open low areas, coastal plains; eastern and central states to Texas.

Bagpods and bequilla are tall annuals, with pinnately compound, alternate leaves and numerous leaflets. Their paperlike pods are about 3 inches long and 3/4-inch wide, some with winged edges. Flowers are yellow and may have red highlights.

All the *Sesbanias* are toxic to horses and livestock, but horses are rarely affected. A saponin may be the toxic principle, with seeds containing the highest amount of poison, and with fall and winter the most dangerous times of year. Symptoms usually appear 1–2 days after ingestion of the plant. Animals may show depression, weakness, diarrhea; liver and kidney failure, ending in death. Animals should be given supportive therapy.

Bunchflower

Melanthium virginicum—Liliaceae, Lily Family—In marshes and wet woodlands of Florida, Texas, and the northeastern and midwestern United States.

Bunchflowers grow to 5 feet tall with long linear leaves originating at the base and smaller leaves close to the green-yellow flowers. Hay containing bunchflowers has caused poisoning. Weakened heartbeat, nausea, slobbering, and sweating are all symptoms. No treatment has been noted.

Fly Poison, or Staggergrass

Amianthium muscaetoxicum—Liliaceae, Lily Family—Woods, fields, and wet areas; in eastern states as well as Oklahoma.

Fly posion is a perennial herb with long, narrow basal leaves and white flowers. The leaves grow in clumps up to 20 inches tall. The flowering stem appears in spring, growing up to 4 feet tall. Its poisonous principle is an alkaloid from the veratrum group, and it is highly toxic to livestock.

Symptoms of plant poisoning include salivation, irregular heartbeat, weakness, and respiratory failure. Most poisonings will occur in spring or summer. There is no treatment for this poisoning and prognosis is poor. Keep animals at pasture well fed, and rotate animals before pastures become overgrazed.

Sleepygrass

Stipa robusta—Poaceae, Grass Family—In woods and throughout the plains of Texas, Colorado, Arizona, and New Mexico.

Sleepygrass grows in clumps up to 4 feet tall with flat 5/16-inch-wide, 2-foot-long leaves. Its branches have small spikelets growing upward. Merely 0.6 percent of an animal's body weight can cause poisoning. The toxic principle is not known. Within 6–24 hours after ingestion, animals become drowsy, depressed, and withdrawn. Pulse and respiration becomes weak and some animals may go down, unable to lift their heads. The sharp grass awns of *Stipa* species can be traumatic to animals' hides. There is no specific treatment. If removed from the source and given supportive therapy, animals usually recover in a few days.

Pictorial Glossary

Shapes and Arrangements of Leaves and Flowers

Heart-shaped Leaf

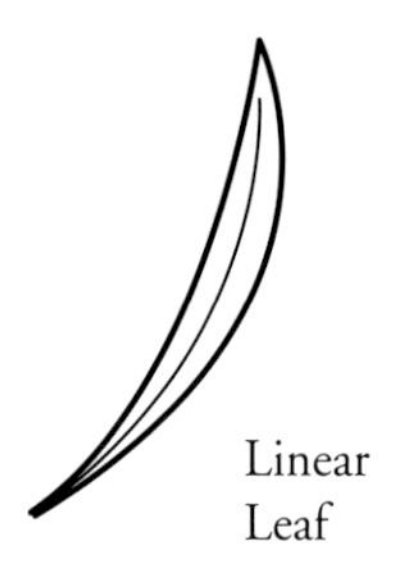
Linear Leaf

Lobed Leaf

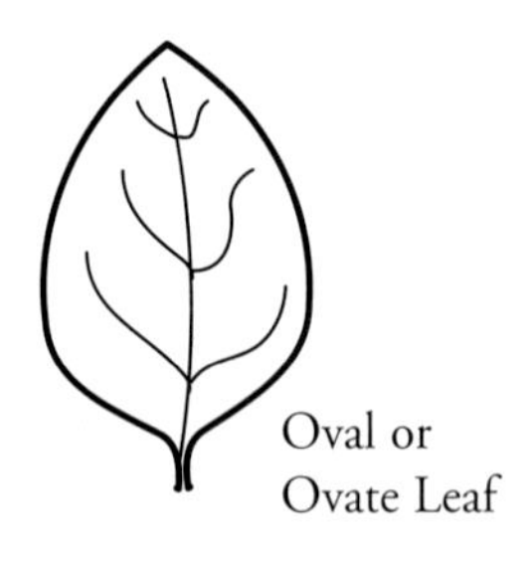
Oval or Ovate Leaf

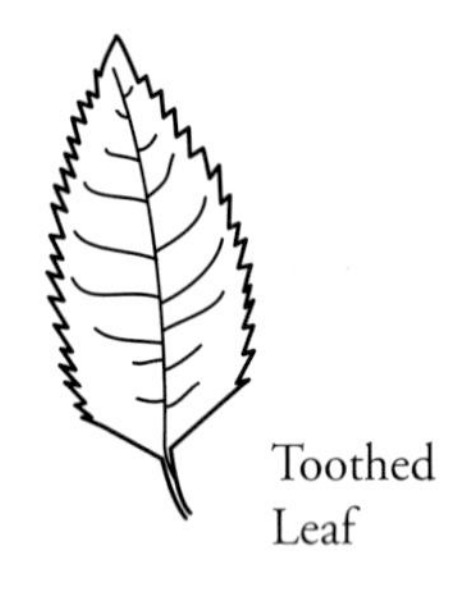
Toothed Leaf

Alternating Leaves

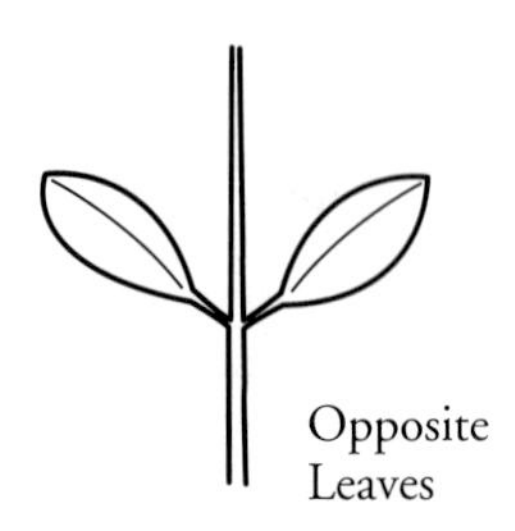
Opposite Leaves

Basal
Leaves

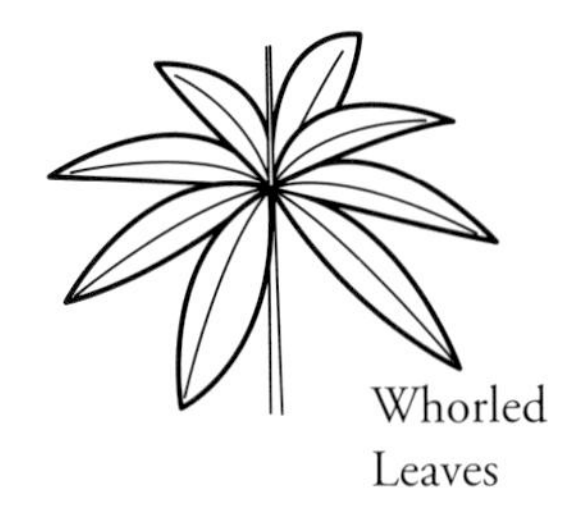
Whorled
Leaves

Compound
Leaf

Palmately
Compound
Leaf

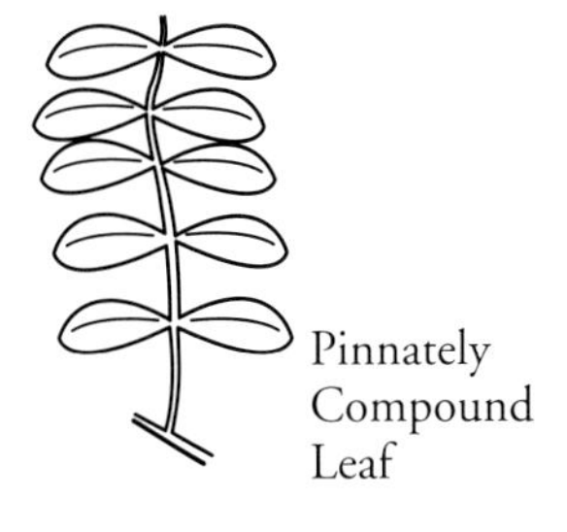
Pinnately
Compound
Leaf

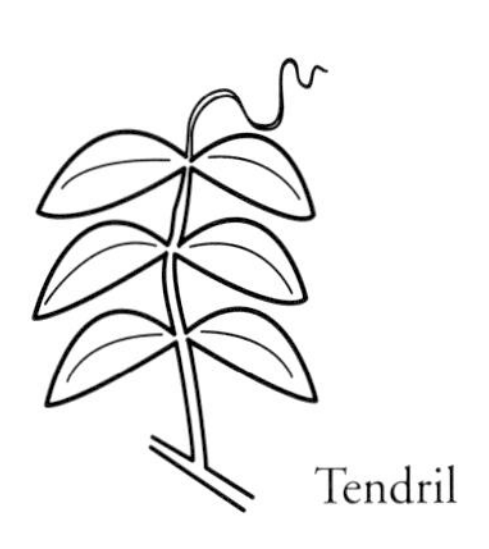
Tendril

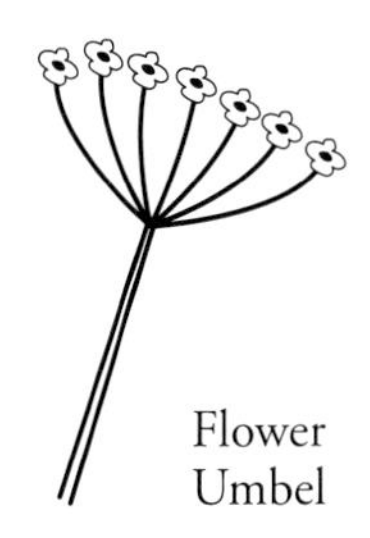
Flower
Umbel

Flower Spike
or Raceme

Bibliography

Bailey, E. Murl, Jr., D.V.M., Ph.D., and Garland, Tam, D.V.M., Ph.D., "Veterinary Toxicology"—lecture and laboratory notes—Texas A&M University, College Station, TX, 1993.

Bailey, Liberty Hyde, and Bailey, Ethel Zoe, *Hortus Third—A Concise Dictionary of Plants Cultivated in the United States and Canada,* revised and expanded by the Staff of the Liberty Hyde Bailey Hortorium–College of Agriculture and Life Sciences, Cornell University; Macmillan Publishing Co. Inc., New York.

Bremness, Lesley, *Herbs*, Dorling Kindersley Publishing, Inc., New York, 1994.

Fraser, Clarence (Editor), *The Merck Veterinary Manual*, Seventh Edition, Merck & Co. Inc., Rahway, NJ, 1991.

Kingsbury, John M., *Poisonous Plants of the U.S. and Canada*, Prentice Hall Inc., Englewood Cliffs, NJ, 1964.

Lampe, Dr. Kenneth F., and McCann, Mary Ann, *AMA Handbook of Poisonous and Injurious Plants*, American Medical Association, Chicago, 1985.

Miller, George, *Landscaping with Native Plants of Texas and the Southwest*, Voyager Press Inc., Stillwater, MN, 1991.

Moore, Jack, "Poisonous Plants: A Survival Guide," *Equus Magazine* #212, June 1995, Equus, Gaithersburg, MD.

Niering, William A., and Olmstead, Nancy C., *Audubon Society Field Guide to North American Wildflowers*, Alfred A. Knopf, New York, 1979.

Siegmund, O.H., *The Merck Veterinary Manual*, Fourth Edition, Merck & Co. Inc., Rahway, NJ, 1973.

Tull, Delena, *A Practical Guide to Edible & Useful Plants*, Texas Monthly Press Inc., Austin, TX, 1987.

Vines, Robert A., *Trees, Shrubs and Woody Vines of the Southwest*, University of Texas Press, Austin, TX, and London, 1960.

Photography Credits

Anderson, E.F., Visuals Unlimited: Fern palm—74.
Anderson, Walt, Visuals Unlimited: Labrador tea flowers—46; Mesquite leaves—58; Winter vetch—134.
Beatty, Bill, Visuals Unlimited: Black walnut flowers—22; Japanese yew fruit—70; Poison hemlock—80.
Bertsch, Jon, Visuals Unlimited: Mojave asters—88.
Bonnie Sue: Red maple—10; Oak—18; Apple—26.
Cavagnaro, D., Visuals Unlimited: Partridge pea—124; Delphinium—166.
Cornell Plantations: Golden chain tree—14; Horsechestnut—20; Mountain laurel—48; Castor bean—54; Milkweed—86; Snakeroot—98; Kale—110; Lupine—130; False hellebore (2)—148; Pokeweed (2)—160; Buckwheat —162; Monkshood (2)—170; Foxglove—172; Rue—188; Horsetails (2)—204.
Cunningham, John D., Visuals Unlimited: Black walnut tree—22; Oleander—34; Saltbush—42; Privet—66; Hydrangea (2)—68; Radish (2)—110; Corn cockle—112; Alsike clover—116; Red clover—120; Partridge pea—124; Cotton pod—156; Indian paintbrush—174; Tobacco (3)—186; Foxtail grass—202.
Ditchburn, Derrick, Visuals Unlimited: Red clover—120; Death camas—146.
Eastman, Don: Black locust tree—16.
Eastman, Priscilla: Yellow star thistle—102; Black locust flowers —16
Eggleston, Patrick M., Visuals Unlimited: Partridge pea—124.
Galati, Carlyn, Visuals Unlimited: Mescal bean—56; Desert aster—88.
Gerlach, Barbara, Visuals Unlimited: Squirrel corn flowers—158.
Gerlach, John, Visuals Unlimited: Goldenbush—96; Tarweed (2)—106; Hairy vetch—134; St. Johnswort—138; Squirrel corn—158; Desert larkspur—166; Indian paintbrush blossom—174.
Gosner, K., Visuals Unlimited: Spreading dogbane—84.
Green, Julie: Peach tree branch—32; Cocklebur—92; Sneezeweed—100; Rattlebox —132; Dallis grass—192; Tall fescue —194; Johnson grass—196.
Gurmankin, A., Visuals Unlimited: Bull nettle—180.
Gustafson, Robert, Visuals Unlimited: Point locoweed—128.
Henley, Mack, Visuals Unlimited : Western labrador tea—46; Poison hemlock flowers—80.

Hill, Arthur R., Visuals Unlimited: Peach tree—32; Cotton blossom—156.

James, Lynn F.: Threadleaf groundsel—94.

Kerstitch, A., Visuals Unlimited: Sticky aster—88.

Knight, A.P.: Rosary pea (2)—60; Groundsel (2)—94; Russian knapweed—102; Hounds tongue (3)—104; Locoweed (3)—126; Sacahuista—152; Jimsonweed (3)—178.

Levine, Hank, Visuals Unlimited: Field of buttercups—164.

Long, M.: Field of cotton in bloom—156.

Loun, George, Visuals Unlimited: Fern palm—74; Golden corydalis—136.

Lyell, Diane C.: Broomweed—90; Buffalo bur —176; Silverleaf nightshade (2)—184.

Lyons, Robert E.: Apples—26; Cherrylaurel —30; Oleander—34; Boxwood—36; Euonymus—38; Kochia—40; Japanese pieris—50; Rhododendron—52; Yellow jessamine—64; Ground hemlock—70; Bracken fern—76; Beefsteak plant—140; Ground ivy—142; Autumn crocus—144; Chives—150; Dutchman's breeches—158; Marsh marigold—168; Arrowgrass (2)—190; Kleingrass—198.

McCutcheon, Steve, Visuals Unlimited: Larkspur—166.

Newman, David, Visuals Unlimited: Mistletoe—154; Foxtail grass—202.

Oliver, Glenn M., Visuals Unlimited: Blackish oxytrope—128; Squirreltail grass (2)—200.

Perkins, Kirtley, Visuals Unlimited: Point locoweed—128.

Prance, G, Visuals Unlimited.: Sudan Grass—196.

Rice, James: Red oak—18; White oak—18; Chinaberry—24; Lantana—72; White sweet clover—122; Onions—150.

Serrao, J., Visuals Unlimited: Red maple leaves—10.

Shiell, Richard: Privet—66; Yellow star thistle—102; Sweet pea—62; Rue flowers—188.

Sieren, David, Visuals Unlimited: Hydrangea—68; Atamasco lily—78.

Sohlden, John, Visuals Unlimited: Water hemlock (2)—82; Hairy vetch—134; St. Johnswort—138.

Sokell, Doug, Visuals Unlimited: Velvet mesquite—58; Spreading dogbane (2)—84.

Spomer, Ron, Visuals Unlimited: Chokecherry fruits—28; Mock azalea (2)—44; Prince's plume (2)—108.

Tatum, Brooking P., Visuals Unlimited: Chokecherry tree and flowers (2)—28; Corydalis—136.

U.S. Department of Agriculture: Leafy spurge (2)—114; Lupine—130

Weber, William J., Visuals Unlimited: Eve's necklace—12; Atamasco lily—78; Crimson clover—118; Showy crotalaria —132; Small rattlebox—132; Corydalis—136; Buttercup blossom—164; Common nightshade—182.

Index

A

Abnormal behavior, 127, 129. *See also:* specific behavior
Abortion
- and Broomweed, 91
- and Dallis Grass, 193
- and False Hellebore, 149
- and Fescue, 195
- and Hairy Vetch, 135
- and Rape, 111
- and Red Clover, 121
- and Red Maple, 11
- and Sweet Clover, 123

Abrin, 61
Abrus precatorius, 60-61
Abscesses, 201
Acer rubrum, 10-11
Aceraceae Family, 10-11
Aconitic acid, 205
Aconitine, 171
Aconitum napellus, 170-171
Acremonium lolii, 193
Aesculin, 21
Aesculus hippocastanum, 20-21
African Rue, 188–189
Aggressive behavior, 95, 107
Agrostemma githago, 112-13
Alertness, 125
Alkaloids
- and African/Mexican Rue, 189
- and Autumn Crocus, 145
- and Boxwood, 37
- and Buffalo Bur, 177
- and Chinaberry, 25
- and Death Camas, 147
- and False Hellebore, 149
- and Fescue, 195
- and Fitweed, 137
- and Foxglove, 173
- and Ground Hemlock, 71
- and Groundsel, 95
- and Horsechestnut, 21
- and Hounds Tongue, 105
- and Jimsonweed, 179
- and Kochia, 41
- and Larkspur, 167
- and Locoweed, 127
- and Lupine, 131
- and Monkshood, 171
- and Poison Hemlock, 81
- and Pokeweed, 161
- and Rattlebox, 133
- and Squirrel Corn, 159
- and Tarweed, 107
- and Yellow Jessamine, 65
- *See also:* Glycoalkaloids; specific alkaloid

Allergic reactions, 23
Allium species, 150-51
Alsike Clover, 116-17
Amaryllidaceae Family, 78-79
Amaryllis Family. *See:* *Amaryllidaceae* Family
American Nightshade. *See:* Common Nightshade
Amines, 155
Aminopropionitrile, 63
Amsinckia intermedia, 106-7
Andromedotoxin. *See:* Grayanotoxin
Anemia
- and Fern Palm, 75
- and Groundsel, 95
- and Hounds Tongue, 105
- and Mesquite, 59
- and Onions, 151
- and Pokeweed, 161
- and Rape, 111
- and Red Maple, 11
- and Sweet Clover, 123
- and Tarweed, 107

Anemonin, 169
Anorexia
- and Apples, 27
- and Broomweed, 91
- and Bull Nettle, 181
- and Cocklebur, 93
- and Coffeeweed, 125
- and Cotton, 157
- and Fescue, 195
- and Hairy Vetch, 135
- and Indian Paintbrush, 175
- and Lantana, 73
- and Oak, 19
- and Sacahuista, 153

Anthraquinone glycoside, 125
Apiaceae Family, 80-83
Apocynaceae Family, 84-85
Apocynum cannabinum, 84-85
Appetite
- and Bracken Fern, 77
- and Foxglove, 173
- and Horsetails, 205
- and Rosary Pea, 61
- and Snakeroot, 99
- *See also:* Anorexia

Apple of Peru (Jimsonweed), 178-179
Apples, 26-27
Apricots, 29
Arabinose, 59
Arrowgrass, 190-191
Asclepiadaceae Family, 86-87
Asclepias species, 86-87
Asphyxiation, 15
Asteraceae Family
  Asters, 88-89
  Broomweed, 90-91
  Cocklebur, 92-93
  Groundsel, 94-95
  Jimmyweed, 96-97
  Snakeroot, 98-99
  Sneezeweed, 100-101
  Yellow Star Thistle, 102-103
Asters, 88-89
Astragalus mollisimus, 126-27
Atamasco Lily, 78-79
Ataxia, 41, 197
Atriplex patula, 42-43
Autumn Crocus, 144-45
Azalea, Mock, 44-45

B

Barley, Wild (Squirreltail Grass), 200-201
Bead tree (Chinaberry), 24-25
Bean
  Castor, 54-55
  Mescal, 56-57
Bean tree (Golden Chain), 14-15
Beargrass (Sacahuista), 152-153
Beech Family. *See: Fagaceae* Family
Beefsteak Plant, 140-141
Belching, 171
Bitterweed (Sneezeweed), 100-101
Black Locust, 16-17
Black Nightshade (Common Nightshade), 182-183
Black Walnut, 22-23
Bladder, 55, 61, 201
Blindness
  and Bracken Fern, 77
  and Buttercup, 165
  and Kochia, 41
  and Rape, 111
  and Red Clover, 121
  and Squirreltail Grass, 201
Bloating
  and Buffalo Bur, 177
  and Cherrylaurel, 31
  and Common Nightshade, 183
  and Dogbane, 85
  and Japanese Pieris, 51
  and Larkspur, 167
  and Milkweed, 87
  and Mock Azalea, 45
  and Monkshood, 171
  and Mountain Laurel, 49
  and Pacific Labrador Tea, 47
  and Poison Hemlock, 81
  and Red Clover, 121
  and Rhododendron, 53
  and Silverleaf Nightshade, 185
  and Water Hemlock, 83
Blood clotting, 123
Blood vessels, constriction of, 193
Bluebonnet (Lupine), 130-131
Borage Family. *See: Boraginaceae* Family
Boraginaceae Family, 104-107
Boxwood, 36-37
Bracken Fern, 76-77
Brassica napus, 110-111
Brassicaceae Family, 108-111
Breathing. *See:* Respiration
Broomweed, 90-91
Buckeye Family. *See: Hippocastinaceae* Family
Buckwheat, 162-163
Buckwheat Family. *See: Polygonaceae* Family
Buffalo Bur, 176-177
Bull Nettle, 180-181
Burning Brush (Kochia), 40-41
Burning Bush, 38-39
Burrow Weed (Jimmyweed), 96-97
Buttercup, 164-165
Buttercup Family. *See: Ranunculaceae* Family
Buxaceae Family, 36-37
Buxus Family. See *Buxaceae* Family
Buxus sempervirens, 36-37

C

California Fern (Poison Hemlock), 80-81
Caltha palustria, 168-169
Caltrop Family. *See: Zygophyllaceae* Family

Carol Bean (Mescal Bean), 56-57
Carolina Jessamine (Yellow Jessamine), 64-65
Carolina Wild Woodbine (Yellow Jessamine), 64-65
Carrot Family. *See: Apiaceae* Family
Caryophyllaceae Family, 112-113
Cassia occidentalis, 124-125
Castilleja species, 174-175
Castor Bean, 54-55
Celastraceae Family, 38-39
Centaurea solstitialis, 102-3
Chenopodiaceae Family, 40-43
Cherry, Wild, 28-29
Cherrylaurel, 30-31
Chewing, 59, 103
"Chewing disease," 103
Chewings Fescue, 194-195
Chinaberry, 24-25
Cicuta maculata, 82-83
Cicutoxin, 83
Circulation, 71
Clover
  Alsike, 116-117
  Crimson, 118-119
  Red, 120-121
  Sweet, 122-123
Coat
  and Asters, 89
  and Broomweed, 91
  and Bull Nettle, 181
  and Fescue, 195
  and Hairy Vetch, 135
  and Indian Paintbrush, 175
  and Oak, 19
  and Prince's Plume, 109
  and Saltbush, 43
Cocklebur, 92-93
Coffee Senna (Coffeeweed), 124-125
Coffeeweed, 124-125
Colchicine, 145
Colchicum autumnale, 144-145
Cold water, sensitivity to, 139
Colic
  and Alsike Clover, 117
  and Apples, 27
  and Black Locust, 17
  and Burning Bush, 39
  and Buttercup, 165
  and Castor Bean, 55
  and Chinaberry, 25
  and Crimson Clover, 119
  and Foxglove, 173
  and Groundsel, 95
  and Japanese Pieris, 51
  and Jimsonweed, 179
  and Larkspur, 167
  and Marsh Marigold, 169
  and Mesquite, 59
  and Mistletoe, 155
  and Mock Azalea, 45
  and Mountain Laurel, 49
  and Oak, 19
  and Pacific Labrador Tea, 47
  and Pokeweed, 161
  and Privet, 67
  and Rape, 111
  and Rhododendron, 53
  and Squirreltail Grass, 201
  and Tarweed, 107
  and Tobacco, 187
  and Water Hemlock, 83
  See also: Gastrointestinal disorders
Collapse
  and Arrowgrass, 193
  and Atamasco Lilly, 79
  and Coffeeweed, 125
  and Fescue, 195
  and Ground Hemlock, 71
  and Japanese Pieris, 51
  and Johnson Grass, 197
  and Mescal Bean, 57
  and Mock Azalea, 45
  and Mountain Laurel, 49
  and Pacific Labrador Tea, 47
Coma
  and Cherrylaurel, 31
  and Death Camas, 147
  and Fern Palm, 75
  and Golden Chain, 15
  and Hydrangea, 69
  and Japanese Pieris, 51
  and Jimsonweed, 179
  and Lupine, 131
  and Mock Azalea, 45
  and Mountain Laurel, 49
  and Oleander, 35
  and Pacific Labrador Tea, 47
  and Peach, 33
  and Poison Hemlock, 81
  and Rhododendron, 53
  and Yellow Jessamine, 65
Common Box (Boxwood), 36-37
Common Nightshade, 182-183

Confusion
and Chinaberry, 25
and Ground Hemlock, 71
See also: Delirium
Coniine, 81
Conium maculatum, 80-81
Constipation
and Broomweed, 91
and Bull Nettle, 181
and Jimmyweed, 97
and Lantana, 73
and Larkspur, 167
and Oak, 19
and Squirreltail Grass, 201
Convulsions
and Arrowgrass, 191
and Castor Bean, 55
and Cherrylaurel, 31
and Chinaberry, 25
and Cocklebur, 93
and Dallis Grass, 193
and Death Camas, 147
and Dogbane, 85
and Fitweed, 137
and Foxglove, 173
and Golden Chain, 15
and Hairy Vetch, 135
and Jimmyweed, 97
and Jimsonweed, 179
and Johnson Grass, 197
and Lupine, 131
and Peach, 33
and Privet, 67
and Sneezeweed, 101
and Squirrel Corn, 159
and Water Hemlock, 83
and Wild Cherry, 29
and Yellow Jessamine, 65
Coordination
and African/Mexican Rue, 189
and Bracken Fern, 77
and Castor Bean, 55
and Golden Chain, 15
and Horsechestnut, 21
and Mescal Bean, 57
and Privet, 67
and Rosary Pea, 61
and Sneezeweed, 101
See also: Staggering; Twitching
Corn Cockle, 112-113
Corydalis species, 136-137
Coryne toxins, 197
Cotton, 156-157
Cough, 135
Cowbane, Spotted (Water Hemlock), 81, 82-83
Cowslip (Marsh Marigold), 168-169
Crabs-Eye (Rosary Pea), 60-61
Creeping Charlie (Ground Ivy), 142-143
Crimson Clover, 118-119
Crocus, Autumn, 144-145
Crotalaria (Rattlebox), 132-133
Crotalaria species, 132-133
Crowfoot. *See:* Buttercup; Monkshood
Cycadaceae Family, 74-75
Cyanide
and Apples, 27
and Apricots, 29
and Arrowgrass, 191
and Cherrylaurel, 31
and Hydrangea, 69
and Johnson Grass, 197
and Peach, 33
and Peach trees, 29
and Wild Cherry, 29
Cycad Family. *See: Cycadaceae* Family
Cycas circinalis, 74-75
Cycasin, 75
Cynoglossum officinale, 104-105
Cytisine, 15, 57

D

Dallis Grass, 192-193
Datura (Jimsonweed), 178-179
Datura species, 178-179
Deadly Nightshade (Common Nightshade), 182-183
Death Camas, 146-147
Deformed offspring, 149, 197
Delirium, 95, 107. *See also*: Confusion
Delphinium species, 166-167
Dennstaedtiaceae Family, 76-77
Depression
and Alsike Clover, 117
and Apples, 27
and Black Walnut, 23
and Bracken Fern, 77
and Buttercup, 165
and Cocklebur, 93
and Fern Palm, 75
and Fitweed, 137
and Groundsel, 95
and Hounds Tongue, 105

and Indian Paintbrush, 175
and Japanese Pieris, 51
and Jimmyweed, 97
and Jimsonweed, 179
and Kochia, 41
and Lupine, 131
and Milkweed, 87
and Mock Azalea, 45
and Mountain Laurel, 49
and Pacific Labrador Tea, 47
and Peach, 33
and Red Maple poisoning, 11
and Rhododendron, 53
and Tarweed, 107
Desert Prince's Plume (Prince's Plume), 108-109
Desert Tobacco, 186-187
Dew poisoning, 117
Diarrhea
and Alsike Clover, 117
and Atamasco Lilly, 79
and Black Locust, 17
and Broomweed, 91
and Burning Bush, 39
and Buttercup, 165
and Castor Bean, 55
and Coffeeweed, 125
and Corn Cockle, 113
and False Hellebore, 149
and Fern Palm, 75
and Fescue, 195
and Ground Hemlock, 71
and Hairy Vetch, 135
and Horsetails, 209
and Hounds Tongue, 105
and Hydrangea, 69
and Jimsonweed, 179
and Marsh Marigold, 169
and Mistletoe, 155
and Oak, 19
and Oleander, 35
and Pokeweed, 161
and Privet, 67
and Rape, 111
and Rattlebox, 133
and Red Clover, 121
and St. Johnswort, 139
and Tobacco, 187
Dicentra canadensis, 158-159
Dicumarol, 123
Digestive system
and Buttercup, 165
and False Hellebore, 149
and Ground Hemlock, 71
and Jimsonweed, 179
and Oak, 19
and Yellow Bristle Grass, 203
Digitalin, 173
Digitalis purpurea, 172-173
Digitoxin, 173
Digoxin, 173
Dogbane, 84-85
Dogbane Family. *See: Apocynaceae* Family
Dullness. *See:* Coat

E

Eagle Fern (Bracken Fern), 76-77
Ears, 193, 195
Edema, 105, 157
Egomaketone, 141
Emaciation, 43, 89, 109, 175, 181
Endophyte fungus, 195
Equisetaceae Family, 204-205
Equisetum arvense, 204-205
Equisetum Family. *See: Equisetaceae* Family
Ergotamine, 193
Ergotoxine, 193
Ericaceae Family
Japanese Pieris, 50-51
Mock Azalea, 44-45
Mountain Laurel, 48-49
Pacific Labrador Tea, 46-47
Rhododendron, 52-53
Esophagus, 177
Euonymus atropurpureus, 38-39
Eupatorium rugosum, 98-99
Euphorbia esula, 114-115
Euphorbiaceae Family, 54-55, 114-115
European Hemlock (Poison Hemlock), 80-81
Eve's Necklace, 12-13
Excitability
and Alsike Clover, 117
and Arrowgrass, 191
and Golden Chain, 15
and Hairy Vetch, 135
and Horsechestnut, 21
and Jimsonweed, 179
and Johnson Grass, 197
Extremities
and Black Locust, 17
and Horsetails, 205
and Oleander, 35
and Poison Hemlock, 81
and Yellow Jessamine, 65

Eyes, 153, 199. *See also*: Blindness; Pupils

F

Fabaceae Family
- Alsike Clover, 116-117
- Black Locust, 17
- Coffeeweed, 124-125
- Crimson Clover, 118-119
- Eve's Necklace, 13
- Golden Chain, 15
- Hairy Vetch, 134-135
- Locoweed, 126-127
- Lupine, 130-131
- Mescal Bean, 56-57
- Mesquite, 58-59
- Oak, 19
- Point Locoweed, 128-129
- Rattlebox, 132-133
- Red Clover, 120-121
- Rosary Pea, 60-61
- Singletary Pea, 62-63
- Sweet Clover, 122-123

Facial muscles, 103
Fagopyrin, 163
Fagopyrum esculentum, 162-63
False Acacia (Black Locust), 16-17
False Hellebore, 148-149
Feces
- and Arrowgrass, 191
- and Boxwood, 37
- and Foxglove, 173
- and Groundsel, 95
- and Johnson Grass, 197
- and Tarweed, 107

Fern
- Bracken, 76-77
- Nebraska or California (Poison Hemlock), 80-81
- Palm, 74-75

Fern Family. *See:* *Dennstaedtiaceae* Family
Fescue
- Chewings, 194–195
- Tall, 194–195

Festuca arundinacea, 194–195
Festuca rubra, 194–195
Fetid Buckeye (Horsechestnut), 20-21
Fetlocks, knuckling of, 97
Fiddleneck (Tarweed), 106-107
Fireball (Kochia), 40-41
Fitweed, 136-137
Foxglove, 172-173
Foxtail Grass. *See:* Squirreltail Grass; Yellow Bristle Grass
Foxtails (Horsetails), 204-205
Frijolito (Mescal Bean), 56-57
Fumariaceae Family, 136-137

G

Gait. *See:* Staggering
Galitoxin resinoid, 87
Gall bladder, 73
Gallotannin, 19
Gangrene, 193, 195
Garget (Pokeweed), 160-161
Gastroenteritis. *See:* Gastrointestinal disorders
Gastrointestinal disorders
- and African and Mexican Rue, 189
- and Autumn Crocus, 145
- and Boxwood, 37
- and Buffalo Bur, 177
- and Cocklebur, 93
- and Common Nightshade, 183
- and Corn Cockle, 113
- and Cotton, 157
- and Fern Palm, 75
- and Hydrangea, 69
- and Lantana, 73
- and Leafy Spurge, 115
- and Milkweed, 87
- and Mistletoe, 155
- and Mock Azalea, 45
- and Mountain Laurel, 49
- and Onions, 151
- and Pacific Labrador Tea, 47
- and Rape, 111
- and Rosary Pea, 61
- and Silverleaf Nightshade, 185
- *See also*: Colic; Intestines

Gelsemine, 65
Gelseminine, 65
Gelsemium sempervirens, 64-65
Gestation, prolonged, 195
Githagenin, 113
Glechoma hederacea, 142-43
Glucosinolates, 111
Glycoalkaloids, 149, 177, 183
Glycosides
- and Apricots, 29
- and Black Locust, 17
- and Cherrylaurel, 31
- and Coffeeweed, 125
- and Fern Palm, 75

and Foxglove, 173
and Horsechestnut, 21
and Hydrangea, 69
and Milkweed, 87
and Oleander, 35
and Peaches, 29
and Privet, 67
and Wild Cherry, 29
Goatweed (St. Johnswort), 138-139
Golden Chain, 14-15
Goldenrod, Rayless (Jimmyweed), 96-97
Goosefoot Family. *See: Chenopodiaceae* Family
Gossypium species, 156-157
Gossypol, 157
Grass
Dallis, 192-193
Fescue, 194-195
Foxtail. *See:* Squirreltail, Yellow Bristle Grass
Johnson, 196-197
Kleingrass, 198-199
Pigeon (Yellow Bristle Grass), 202-203
Rye, 193
Squirreltail, 200-201
Yellow Bristle, 202-203
Grass Family. *See: Poaceae* Family
Grayanotoxins
and Japanese Pieris, 51
and Mock Azalea, 45
and Mountain Laurel, 49
and Pacific Labrador Tea, 47
and Rhododendron, 53
Great Laurel (Rhododendron), 52-53
Ground Hemlock, 70-71
Ground Ivy, 142-143
Groundsel, 94-95
Growth rate, 157, 195
Guitierrezia microcephala, 90-91

H

Hair loss, 43, 89, 115, 135, 175
Hairy Vetch, 134-135
Halostachine, 195
Haplopappus heterophyllus, 96-97
Harmaline, 189
Harmalol, 189
Harmine, 189
Heart
and Asters, 89
and Buffalo Bur, 177
and Common Nightshade, 183
and Cotton, 157
and False Hellebore, 149
and Jimmyweed, 97
and Larkspur, 167
and Oleander, 35
and Prince's Plume, 109
and Silverleaf Nightshade, 185
Heartbeat
and Cherrylaurel, 31
and Cocklebur, 93
and Death Camas, 147
and Foxglove, 173
and Monkshood, 171
and Peach, 33
and Poison Hemlock, 81
and St. Johnswort, 139
and Sweet Clover, 123
See also: Pulse
Heath Family. *See: Ericaceae* Family
Helanin, 101
Helenium species, 100-101
Hemlock
Ground, 70-71
Poison, 80-81
Water, 81, 82-83
Hemorrhages
and African and Mexican Rue, 189
and Castor Bean, 55
and Fern Palm, 75
and Snakeroot, 99
and Sweet Clover, 123
Herba mora negra (Common Nightshade), 182-183
Hippocastinaceae Family, 20-21
Hog Brake (Bracken Fern), 76-77
Honey dew fungus, 193
Honey Mesquite (Mesquite), 58-59
Hooves
and Alsike Clover, 117
and Asters, 89
and Fescue, 195
and Indian Paintbrush, 175
and Prince's Plume, 109
and Saltbush, 43
gangrene of, 195
Hordeum jubatum, 200–201

Horse Nettle (Bull Nettle), 180-181
Horsechestnut, 20-21
Horsetails, 204-205
Hounds Tongue, 104-5
Hydrangea, 68-69
Hydrangea species, 68-69
Hydrocyanic acid. *See:* Prussic acid
Hyoscine, 179
Hyoscyamine, 179
Hypericaceae Family, 138-39
Hypericin, 139
Hypericum perforatum, 138-39

I, J, K

Indian Hemp (Dogbane), 84-85
Indian Lilac (Chinaberry), 24-25
Indian Paintbrush, 174-175
Indian Poke (False Hellebore), 148-149
Infertility, 195
Intestines, 55, 59, 119, 181. *See also*: Gastrointestinal disorders
Isoegomaketone, 141
Isoguinoline, 159
Isoquinoline, 137
Ivy, Ground, 142-143
Jamestownweed (Jimsonweed), 178-179
Jaundice
- and Alsike Clover, 117
- and Castor Bean, 55
- and Groundsel, 95
- and Hounds Tongue, 105
- and Rape, 111
- Tarweed, 107

Jequirity Bean (Rosary Pea), 60-61
Jessamine, Yellow, 64-65
Jimmy Goldenweed (Jimmyweed), 96-97
Jimmyweed, 96-97
Jimsonweed, 178-179
Johnson Grass, 196-197
Juglandaceae Family, 22-23
Juglans species, 22-23
Juglone, and Black Walnut, 23
Juncaginaceae Family, 190-191
Kalmia latifolia, 48-49
Kansas Thistle (Buffalo Bur), 176-177
Kidneys
- and Broomweed, 91
- and Coffeeweed, 125
- and Fescue, 195
- and Hairy Vetch, 135
- and Japanese Pieris, 51
- and Jimmyweed, 97
- and Oak, 19
- and Onions, 151
- and Rape, 111
- and Rhododendron, 53
- and Rosary Pea, 61
- and Sacahuista, 153
- and Snakeroot, 99

Klamathweed (St. Johnswort), 138-139
Kleingrass, 198-199
Kochia, 40-41
Kochia scoparia, 40-41

L

Labrador Tea, Pacific, 46-47
Laburnum anagyroides, 14-15
Lactones, 101
Lameness
- and Asters, 89
- and Dallis Grass, 193
- and Fescue, 195
- and Indian Paintbrush, 175
- and Prince's Plume, 109
- and Saltbush, 43
- and Sweet Clover, 123

Lamiaceae Family, 140-143
Laminitis, 23, 63
Lantadene A and B, 73
Lantana, 72-73
Lantana camara, 72-73
Larkspur, 166-167
Lathyrus species, 62-63
Laurel, Mountain, 48-49
Leafy Spurge, 114-115
Ledum columbianum, 46-47
Legs, 23, 159, 197
Ligustrin, 67
Ligustron, 67
Ligustrum (Privet), 66-67
Ligustrum vulgare, 66-67
Liliaceae Family
- Autumn Crocus, 144-145
- Death Camas, 146-147
- False Hellebore, 148-149
- Onions, 150-151
- Sacahuista, 152-153

Lily, Atamasco, 78-79
Lily Family. *See: Liliaceae* Family

Lips, 103, 115, 169
Liver
and African and Mexican Rue, 189
and Alsike Clover, 117
and Asters, 89
and Broomweed, 91
and Buffalo Bur, 177
and Cocklebur, 93
and Coffeeweed, 125
and Fern Palm, 75
and Fescue, 195
and Groundsel, 95
and Hairy Vetch, 135
and Hounds Tongue, 105
and Japanese Pieris, 51
and Jimmyweed, 97
and Kleingrass, 199
and Lantana, 73
and Lupine, 131
and Mock Azalea, 45
and Mountain Laurel, 49
and Oak, 19
and Onions, 151
and Pacific Labrador Tea, 47
and Prince's Plume, 109
and Rape, 111
and Rhododendron, 53
and Rosary Pea, 61
and Sacahuista, 153
and Snakeroot, 99
and Tarweed, 107
Locoweed, 126-127, 128-129
Locust, Black, 17
Logania Family. *See: Loganiaceae* Family
Loganiaceae Family, 64-65
Loranthaceae Family, 154-155
Lungs, 61, 177, 183, 185
Lupine, 130-131
Lupinus species, 130-131
Lysergic acid, 193

M

Machaeranthera species, 88-89
Mallow Family. *See: Malvaceae* Family
Malus sylvestris, 26-27
Malvaceae Family, 156-157
Mane, 43, 89, 109, 175
Maple Family. See *Aceraceae* Family
Maple, Red, 11
Marigold, Marsh, 168-169
Melia azedarach, 24-25
Meliaceae Family, 24-25
Melilotus species, 122-123
Menziesia ferruginea, 44-45
Mescal Bean, 56-57
Mesquite, 58-59
Mexican Fireweed (Kochia), 40-41
Mexican Rue, 189
Milk, mare's, 195
"Milk sickness," 99
Milkweed, 86-87
Milkweed Family. *See: Asclepiadaceae* Family
Mint Family. *See: Lamiaceae* Family
Mistletoe, 154-155
Mistletoe Family. *See: Loranthaceae* Family
Mock Azalea, 44-45
Mock-Orange (Cherrylaurel), 30-31
Monkshood, 170-171
Monocrotaline, 133
Mouth, 101, 103, 177, 181, 203. *See also*: Lips; Tongue
Mucous membranes
and Apples, 27
and Arrowgrass, 191
and Cherrylaurel, 31
and Johnson Grass, 197
and Peach, 33
and Red Maple, 11
and Wild Cherry, 29
and Yellow Bristle Grass, 203
Muscles, 41. *See also*: Collapse; Rigidity; Stiffness; Trembling; Twitching; Weakness
Musquash Root (Water Hemlock), 81, 82-83
Mustard Family. *See: Brassicaceae* Family
Muzzle, 19, 91, 117, 167

N

N-methylmorpholin, 125
N-propyl disulfide, 151
Nasal discharge
and Broomweed, 91
and Hairy Vetch, 135
and Japanese Pieris, 51
and Mock Azalea, 45
and Mountain Laurel, 49
and Pacific Labrador Tea, 47

and Sacahuista, 153
Nausea
and Cocklebur, 93
and Death Camas, 147
and Fern Palm, 75
and Japanese Pieris, 51
and Mock Azalea, 45
and Mountain Laurel, 49
and Pacific Labrador Tea, 47
and Rhododendron, 53
and Snakeroot, 99
Nebraska Fern (Poison Hemlock), 80-81
Necrosis, 19
Nerium Oleander, 34-35
Nervous system
and Alsike Clover, 117
and Common Nightshade, 183
and Dallis Grass, 193
and Horsechestnut, 21
and Hounds Tongue, 105
and Jimsonweed, 179
and Johnson Grass, 197
and Locoweed, 127
and Silverleaf Nightshade, 185
See also: Bloating; Tiredness; Nervousness; Trembling
Nervousness
and Buttercup, 165
and Dallis Grass, 193
and Ground Hemlock, 71
and Groundsel, 95
and Lupine, 131
and Marsh Marigold, 169
and Mesquite, 59
and Poison Hemlock, 81
and Tarweed, 107
Nettle, Bull, 180–181
Nicotiana attenuata, 186-187
Nicotiana glauca, 186-187
Nicotiana species, 186-187
Nicotiana trigonophylla, 186-187
Nicotine, 187
Nightshade
Common, 182–183
Silverleaf, 184-185
Nightshade Family. *See: Solanaceae* Family
Nigropallidal encephalomalacia, 103
Nitrates, 41, 197
Nitroamines, 127
Nitrogen oxides, 131
Nolina texana, 152-153

O, P

Oak, 19
Oleaceae Family, 66-67
Oleander, 34-35
Onions, 150-151
Orach (Saltbush), 42-43
Oxalates, 41
Oxytropis species, 128-129
Pacific Labrador Tea, 46-147
Pain, acute, 159, 173
Palm, Fern, 74-75
Palustrine, 205
Panicum coloratum, 198-199
Papaveraceae Family, 158-159
Paralysis
and Black Locust, 17
and False Hellebore, 149
and Fern Palm, 75
and Jimmyweed, 97
and Jimsonweed, 179
and Mescal Bean, 57
and Oleander, 35
and Peach, 33
and Poison Hemlock, 81
and Silverleaf Nightshade, 185
Paspalum dilatatum, 192-193
Pea
Rosary, 60-61
Singletary, 62-63
Pea Family. *See: Fabaceae* Family
Peaches, 29, 32-33
Peganum harmala, 189–190
Perilla frutescens, 140-141
Perilla ketone, 141
Perilla Mint. *See:* Beefsteak Plant
Perloline, 195
Phasin, 17
Phoradendron villosum, 154-155
Photosensitivity
and Alsike Clover, 117
and Atamasco Lilly, 79
and Buckwheat, 163
and Crimson Clover, 119
and Groundsel, 95
and Hairy Vetch, 135
and Hounds Tongue, 105
and Kleingrass, 199
and Kochia, 41
and Lantana, 73
and Leafy Spurge, 115
and Rape, 111
and Sacahuista, 153
and St. Johnswort, 139

and Tarweed, 107
Phototoxins, 117
Phytolacca americana, 160-161
Phytolaccaceae Family, 160-161
Phytolaccatoxin, 161
Phytolaccin, 161
Phytotoxins, 17, 61
Pieris floribunda, 51
Pieris, Japanese, 50-51
Pieris japonica, 50-51
Pigeon Berry (Pokeweed), 160-161
Pigeon Grass (Yellow Bristle Grass), 202-203
Pink Family. *See:* *Caryophyllaceae* Family
Piperidine, 131
Placentas, retained, 195
Pneumonia, 141
Poaceae Family
Dallis Grass, 192-193
Fescue, 194-195
Johnson Grass, 196-197
Kleingrass, 198-199
Squirreltail Grass, 200-201
Yellow Bristle Grass, 202-203
Point Locoweed, 128-129
Poison Hemlock, 80-81
Poisonweed (Larkspur), 166-167
Pokeweed, 160-161
Pokeweed Family. *See:* *Phytolaccaceae* Family
Polycyclic diterpenoid alkaloids, 167
Polygonaceae Family, 162-163
Pot belly, 157
Precatory Bean (Rosary Pea), 60-61
Prince's Plume, 108-109
Privet, 66-67
Prosopis glandulosa, 58-59
Proteins, toxic, 155
Protoanemonin, 165
Prunus laurocerasus, 30-31
Prunus persica, 32-33
Prunus species, 28-29
Prussic acid, 33, 191, 197
Pteridium aquilinum, 76-77
Pulse
and Arrowgrass, 191
and Black Locust, 17
and Castor Bean, 55
and Dogbane, 85
and Foxglove, 173
and Horsetails, 205
and Jimmyweed, 97
and Johnson Grass, 197
and Larkspur, 167
and Lupine, 131
and Milkweed, 87
and Privet, 67
and Tobacco, 187
and Wild Cherry, 29
See also: Heartbeat
Pupils
and Black Locust, 17
and Ground Ivy, 143
and Jimsonweed, 179
and Poison Hemlock, 81
and Water Hemlock, 83
Purple Mint (Beefsteak Plant), 140-41
Pyrrolizidine, 95, 105, 107, 133

Q, R

Quercus species, 18-19
Quinolizidine, 131
Ranunculaceae Family
Buttercup, 164-165
Larkspur, 166-167
Marsh Marigold, 168-169
Monkshood, 170-171
Ranunculin, 165
Ranunculus species, 164-165
Rape, 110-111
Rattlebox, 132-133
Rayless Goldenrod (Jimmyweed), 96-97
Rectum, 197
Red blood cells, 11, 151
Red Clover, 120-121
Reflexes, 103
Respiration
and Apples, 27
and Arrowgrass, 191
and Beefsteak Plant, 141
and Black Walnut, 23
and Boxwood, 37
and Cherrylaurel, 31
and False Hellebore, 149
and Fitweed, 137
and Ground Ivy, 143
and Hydrangea, 69
and Jimsonweed, 179
and Johnson Grass, 197
and Larkspur, 167
and Lupine, 131
and Mescal Bean, 57
and Milkweed, 87

and Peach, 33
and Poison Hemlock, 81
and Pokeweed, 161
and Silverleaf Nightshade, 185
and Squirrel Corn, 159
and Tobacco, 187
and Wild Cherry, 29
and Yellow Jessamine, 65
Restlessness, 171
Rhododendron, 52-53
Rhododendron maximum, 52-53
Richweed (Snakeroot), 98-99
Ricin, 55
Ricinus communis, 54-55
Rigidity, 189, 205
Robin, 17
Robinia pseudoacacia, 16-17
Robitin, 17
Rosaceae Family
and Apples, 26-27
and Cherrylaurel, 30-31
and Peach, 32-33
and Wild Cherry, 28-29
Rosary Pea, 60-61
Rose Family. *See: Rosaceae* Family
Rosebay (Rhododendron), 52-53
Rue, African, 188–189
Rue, Mexican, 189
Rustyleaf (Mock Azalea), 44-45
Rye Grass, 195

S

Sacahuista, 152-53
St. Johnswort, 138-39
St. Johnswort Family. *See: Hypericaceae* Family
Salivation
and African and Mexican Rue, 189
and Arrowgrass, 191
and Buttercup, 165
and Death Camas, 147
and Ground Ivy, 143
and Hairy Vetch, 135
and Japanese Pieris, 51
and Johnson Grass, 197
and Marsh Marigold, 169
and Mesquite, 59
and Mock Azalea, 45
and Monkshood, 171
and Mountain Laurel, 49
and Pacific Labrador Tea, 47
and Pokeweed, 161
and Red Clover, 121
and Rhododendron, 53
and Snakeroot, 99
and Water Hemlock, 83
and Wild Cherry, 29
Saltbush, 42-43
Sand burn, 79
Sapogenin, 113
Saponins
and Broomweed, 91
and Chinaberry, 25
and Foxglove, 173
and Horsechestnut, 21
and Kleingrass, 199
and Pokeweed, 161
Saxifragaceae Family, 68-69
Saxifrage Family. *See: Saxifragaceae* Family
Scoke (Pokeweed), 160-161
Scouring Rushes (Horsetails), 204-205
Scrophulariaceae Family, 172-175
Seizures, 87, 137
Selenium
and Asters, 89
and Indian Paintbrush, 175
and Locoweed, 127
and Prince's Plume, 109
and Saltbush, 43
Senecio species, 94-95
Senna (Coffeeweed), 124-125
Sesquiterpene lactones, 101
Setaria Lutescens, 202-203
Silverleaf Nightshade, 184-185
Skin
and Alsike Clover, 117
and Buckwheat, 163
and Leafy Spurge, 115
and Marsh Marigold, 169
and St. Johnswort, 139
and Squirreltail Grass, 201
Skunk Cabbage (False Hellebore), 148-149
Slaframine, 121
Slobbering. *See:* Salivation
Sluggishness. *See:* Tiredness
Snakeroot, 98-99
Snakeweed, 91
Snapdragon Family. *See: Scrophulariaceae* Family
Snapping, 137
Sneezeweed, 100-101
Solanaceae Family
Buffalo Bur, 176-177
Bull Nettle, 180-181

Common Nightshade, 182-183
Jimsonweed, 178-179
Silverleaf Nightshade, 184-185
Tobacco, 186-187
Solanine
and Buffalo Bur, 177
and Bull Nettle, 181
and Common Nightshade, 183
and Jimsonweed, 179
and Silverleaf Nightshade, 185
Solanum americanum, 182-183
Solanum carolinense, 180-181
Solanum elaeagnifolium, 184-185
Solanum rostratum, 176-177
Sophora affinis, 12-13
Sophora secundiflora, 56-57
Sorghum halepense, 196-197
Spleen, 177, 183, 185
Spotted Cowbane (Water Hemlock), 81, 82-83
Spotted Hemlock (Poison Hemlock), 80-81
Spurge Family. *See: Euphorbiaceae* Family
Squirrel Corn, 158-159
Squirreltail Grass, 201-202
Staff Tree Family. *See: Celastraceae* Family
Staggergrass (Squirrel Corn), 158-159
Staggering
and Apples, 27
and Arrowgrass, 191
and Asters, 89
and Atamasco Lilly, 79
and Cherrylaurel, 31
and Coffeeweed, 125
and Dallis Grass, 193
and Death Camas, 147
and Dogbane, 85
and Fescue, 195
and Fitweed, 137
and Horsetails, 205
and Indian Paintbrush, 175
and Johnson Grass, 197
and Milkweed, 87
and Peach, 33
and Prince's Plume, 109
and Singletary Pea, 63
Staggerweed (Larkspur), 166-167
Staggerwort (Sneezeweed), 100-101
Stanleya pinnata, 108-109
Star Thistle, Yellow, 102-103
Stiffness
and Asters, 89
and Hairy Vetch, 135
and Indian Paintbrush, 175
and Jimmyweed, 97
and Larkspur, 167
and Mescal Bean, 57
and Prince's Plume, 109
and Red Clover, 121
and Singletary Pea, 63
Stillbirth, 195
Stinking Buckeye (Horse-chestnut), 20-21
Stomach, 55, 157. *See also*: Colic; Gastrointestinal disorders
Strychnine, 65
Stupor, 133
Sulfates, 41
Summer Cypress (Kochia), 40-41
Sunflower Family. *See: Asteraceae* Family
Sunflower, Swamp (Sneezeweed), 100-101
Swainsonine, 127
Swallowing
and Japanese Pieris, 51
and Mock Azalea, 45
and Monkshood, 171
and Mountain Laurel, 49
and Pacific Labrador Tea, 47
and Rhododendron, 53
Swamp Sunflower (Sneezeweed), 100-101
Sweating, 55, 143
Sweet Clover, 122-123
Swelling, 123, 135, 157, 163

T

Tachycardia, 193, 201
Tail, 43, 89, 109, 175, 193, 195
Tannin, 19
Tarweed, 106-107
Taxaceae Family, 70-71
Taxine, 71
Taxus canadensis, 70-71
Teeth grinding
and Japanese Pieris, 51
and Mock Azalea, 45
and Mountain Laurel, 49

and Pacific Labrador Tea, 47
and Water Hemlock, 83
Temperature
and Dogbane, 85
and Fescue, 195
and Jimsonweed, 179
and Milkweed, 87
and Rosary Pea, 61
and St. Johnswort, 139
and Water Hemlock, 83
Texas Mountain Laurel (Mescal Bean), 56-57
Texas Sophora (Eve's Necklace), 12-13
Texas Thistle (Buffalo Bur), 176-177
Texas Umbrella (Chinaberry), 24-25
Thiaminase, 77, 209
Thirst, 19, 181
Thorn Apple (Jimsonweed), 178-179
Thyroid, 111
Tipton Weed (St. Johnswort), 138-139
Tiredness
and Buffalo Bur, 177
and Common Nightshade, 183
and Groundsel, 95
and Horsechestnut, 21
and Silverleaf Nightshade, 185
and Tarweed, 107
Tobacco, 186-187
Tongue, 115, 117
Tree Tobacco, 186-187
Trembling
and African and Mexican Rue, 189
and Buffalo Bur, 177
and Castor Bean, 55
and Dallis Grass, 193
and Fescue, 195
and Horsetails, 205
and Jimmyweed, 97
and Mescal Bean, 57
and Mesquite, 59
and Oleander, 35
and Poison Hemlock, 81
and Rye Grass, 193
and Snakeroot, 99
and Squirrel Corn, 159
and Water Hemlock, 83
Tremetol, 97
Trifoliosis, 117
Trifolium hybridum, 116-117
Trifolium incarnatum, 118-119
Trifolium pratense, 120-121
Triglochin maritima, 190-191
Triglochin palustris, 190-191
Trompillo (Silverleaf Nightshade), 184-185
Tumors, 75
Twitching
and Common Nightshade, 183
and Fitweed, 137
and Horsechestnut, 21
and Larkspur, 167
and Silverleaf Nightshade, 185
and Tobacco, 187

U, V

Ulcers
and Buckwheat, 163
and Buttercup, 165
and Marsh Marigold, 169
and Oak, 19
and Pokeweed, 161
and Rosary Pea, 61
and Squirreltail Grass, 201
and Yellow Bristle Grass, 203
Urinary tract, 197
Urine
and African and Mexican Rue, 189
and Arrowgrass, 191
and Buttercup, 165
and Castor Bean, 55
and Coffeeweed, 125
and Foxglove, 173
and Groundsel, 95
and Jimmyweed, 97
and Johnson Grass, 197
and Oak, 19
and Red Maple, 11
and Sacahuista, 153
and Tarweed, 107
Vasicine, 189
Veratrum alkaloids, 147
Veratrum viride, 148-149
Verbena Family. *See: Verbenaceae* Family
Verbenaceae Family, 72-73
Vetch. *See:* Hairy Vetch
Vicia villosa, 134-135
Volatile oils, 143

W

Walking
aimless, 95, 107, 175
in circles, 133
and Groundsel, 95
and Indian Paintbrush, 175
and Rattlebox, 133
and Singletary Pea, 63
and Tarweed, 107
Walnut, Black, 22-23
Walnut Family. *See: Juglandaceae* Family
Water Hemlock, 81, 82-83
Weakness
and Alsike Clover, 117
and Apples, 27
and Black Locust, 17
and Castor Bean, 55
and Cocklebur, 93
and Death Camas, 147
and Hairy Vetch, 135
and Horsetails, 205
and Hydrangea, 69
and Jimmyweed, 97
and Lantana, 73
and Milkweed, 87
and Monkshood, 171
and Pokeweed, 161
and Rattlebox, 133
and Red Maple poisoning, 11
and Rhododendron, 53
and Sacahuista, 153
and Sneezeweed, 101
and Sweet Clover, 123
and Yellow Jessamine, 65
Weight
and Alsike Clover, 117
and Fescue, 195
and Groundsel, 95
and Hounds Tongue, 105
and Locoweed, 127
and Point Locoweed, 129
and Rape, 111
and Tarweed, 107
White Horse Nettle (Silverleaf Nightshade), 184-185
White Muscle Disease, 125
White Sanide (Snakeroot), 98-99
Wild Barley (Squirreltail Grass), 200–201
Wild Cherry, 28-29
Wild Coffee (Coffeeweed), 124-125
Wild Tobacco, 186-187
Winter Vetch (Hairy Vetch), 134-135

X,Y,Z

Xanthium species, 92-93
Yellow Bristle Grass, 202-203
Yellow Jessamine, 64-65
Yellow Star Thistle, 102-103
Yew Family. *See: Taxaceae* Family
Zephyranthes atamasco, 78-79
Zigadenus species, 146-147
Zygophyllaceae Family, 188-189